TEXTBOOK OF MATERIALS AND METALLURGICAL THERMODYNAMICS

Textbook of Materials and Metallurgical Thermodynamics

AHINDRA GHOSH
Formerly Professor
Materials and Metallurgical Engineering
Indian Institute of Technology Kanpur

PHI Learning Private Limited
Delhi-110092
2015

₹ 250.00

TEXTBOOK OF MATERIALS AND METALLURGICAL THERMODYNAMICS
Ahindra Ghosh

ISBN-978-81-203-2091-8

The export rights of this book are vested solely with the publisher.

Seventh Printing ... ... **September, 2015**

Published by Asoke K. Ghosh, PHI Learning Private Limited, Rimjhim House, 111, Patparganj Industrial Estate, Delhi-110092 and Printed by V.K. Batra at Pearl Offset Press Private Limited, New Delhi-110015.

To

(Late) Shri Sajoy Panchanan Ghosh
(Late) Shri Gobinda Lal Sarkar
Shrimati Binapani Biswas

Contents

Preface

Metallurgical thermodynamics, as also its modified version, 'Thermodynamics of Materials', constitutes one of the principal scientific foundations of the disciplines of Metallurgical Engineering and Materials Science and Engineering. Hence, all metallurgical and materials science departments teach it as a compulsory course in their Bachelor's degree programme.

There are only a few books published so far, which are suited as textbooks for such a course. These foreign-authored books are well-written, but somewhat voluminous and expensive. I have taught this subject at the undergraduate level at the Indian Institute of Technology Kanpur for several years. My experience is that both students and teachers would be considerably benefited if a more concise text is available in the market. It will make teaching as well as understanding of the basics better. Moreover, students will be able to afford it for personal acquisition.

I started writing this book with these objectives in mind. The book contains essential figures and tables, solved examples and problems to illustrate the concepts discussed. Data for solving problems as well as answers to problems have also been provided.

The laws of thermodynamics are general. However, their applications are specific to various disciplines. That is why a general book on thermodynamics does not meet the requirements of all branches of science and engineering, and specialized texts are required. We may classify thermodynamics into:

(a) *Thermodynamics of Non-reactive Systems*: Branches such as physics and mechanical engineering are primarily concerned with this; books on Engineering Thermodynamics have major coverage on this.

(b) *Thermodynamics of Reactive Systems* (popularly called *Chemical Thermodynamics*): Here, physico-chemical processes and chemical reactions are primarily dealt with, and the Gibbs Free Energy criterion is extensively employed for equilibrium calculations, thermodynamic properties of phases and solutions; calculation of heats and entropies of reactions, and physico-chemical processes are other important features.

Broadly speaking, metallurgical/materials thermodynamics belongs to chemical thermodynamics. However, the major emphasis is on high temperature systems and processes involving solid or liquid solutions of metals and inorganic compounds. For this, large number of experimental thermodynamic measurements have been carried out in the last 6–7 decades, and some special relations and procedures have also been evolved to deal with these solutions. Hence, the subject attained maturity and became an Engineering Science in its own merit only from the decade of 1960 or so, although the general thermodynamic laws and relations were well-established by 1930s.

In the beginning of the 20th Century, metallurgy was primarily an art. Application of pure and engineering sciences gave it recognition as a branch of engineering by 1950s, and contributed to the development of metallurgical sciences, including metallurgical thermodynamics. Atomic and solid state physics also matured by 1930s/1940s. In the decade of 1940s, major advances were also initiated in high technology areas such as nuclear, space and aviation technologies, and solid state electronics. These required a variety of new special purpose materials. Metallurgical sciences, solid state physics etc. provided scientific foundation for the same. Earlier, metallurgical engineers were primarily concerned with metals and alloys, and ceramists with traditional oxide ceramics, such as refractories. However, the new materials, both metallic and non-metallic, changed the character of academic programmes. It was realized that the fundamentals of all these are the same, and hence can be taught in a generalized way. This led to the growth of Materials Science and Engineering.

In the IITs and engineering colleges in India, Bachelor's degree programme is offered in metallurgical engineering or in a combined programme of metallurgical and materials engineering. This text aims to cater to the basic one-semester undergraduate course on thermodynamics. There are examples and problems on both metallic and inorganic non-metallic materials. Chapter 14 has an additional brief presentation on thermodynamics of surfaces as well as interfaces and defects in solids of importance to materials science in general.

Chapters 1–5 cover general thermodynamics and thermochemistry. Chapter 12 presents some introductory discussions on statistical thermodynamics. The remaining chapters (i.e. Chapters 6–11 and 13) deal with standard topics of chemical and metallurgical/materials thermodynamics. As stated earlier, the principal emphasis is on high temperature systems and processes of interest to metallurgy and materials science.

I wish to express my gratitude to the All India Council for Technical Education (AICTE) for providing Emeritus Fellowship and miscellaneous financial assistance during preparation of the manuscript. I am also indebted to the Indian Institute of Technology Kanpur for making infrastructural facilities available, which has made this work possible. I wish to thank Mr. J.L. Kuril for typing the manuscript, Mr. B.K. Jain for tracing the diagrams, and Mr. A. Sharma for assistance in many ways. Finally, I sincerely thank my wife, Radha and other family members without whose cooperation and encouragement the writing of this book would not have been possible.

AHINDRA GHOSH

List of Symbols with Units

("–" indicates a dimensionless quantity; mol means gram mole (i.e. mole); J means Joule)

A	Helmholtz free energy, surface area	J or J mol^{-1}; m^2
a_i	activity of component i in a solution	–
C	number of components in a system	–
C_V, C_P	heat capacities at constant volume and constant pressure, respectively	J mol^{-1} K^{-1}
E	EMF of an electrolytic cell	volt
e_i^j	interaction coefficient describing influence of solute j on f_i, in 1 wt.% standard state	–
f_i	fugacity of component i in a solution; also activity coefficient of i in 1 wt.% standard state	–
F	Faraday's constant	J volt^{-1} g–equi^{-1}
G	Gibbs free energy	J, J mol^{-1}
G^0	Gibbs free energy at standard state	J, J mol^{-1}
ΔG	change of G due to a finite process	J, J mol^{-1}
H	enthalpy	J, J mol^{-1}
h	Planck's constant	J·S
h_i	activity of component i in a solution in 1 wt. % standard state	–
i	general symbol of a component in a solution	–
J	activity quotient	–
K	equilibrium constant	–
k_B	Boltzmann's constant	J K^{-1} molecule^{-1}
K_h	equilibrium constant in 1 wt. % standard state	–
M	molecular/atomic mass	g mol^{-1}

xvii

m	mass of one molecule/atom	g molecule^{-1}
n	number of gram moles	–
N	number of molecules/atoms	–
N_0	Avogadro's number	
P P_T	total pressure in a system	atm
P	number of phases	–
p_i	partial pressure of component i in a gas mixture	atm
q	quantity of heat transferred	J
Q	general symbol of an extensive state property of a solution (G, H, U, S, V, etc.) per mole	J mol^{-1}, m^3 mol^{-1}, etc.
Q'	value of Q for the entire solution	J, m^3, etc.
$\bar{Q}_i$	partial molar value of Q in a solution (e.g. $\bar{G}_i, \bar{H}_i$)	J mol^{-1}, etc.
$\bar{Q}_i^m$	partial molar value for mixing of Q in a solution (e.g. $\bar{G}_i^m, \bar{H}_i^m, \bar{S}_i^m$, etc.)	J mol^{-1}, etc.
ΔQ^m	integral molar value of Q for mixing, i.e. change of Q upon mixing (e.g. $\Delta S^m, \Delta G^m$, etc.)	J mol^{-1}, etc.
Q^{XS}	excess molar value of Q (e.g. G^{XS}, etc.)	J mol^{-1}, etc.
R	Universal gas constant	J mol^{-1}K^{-1}, etc.
S	entropy	J K^{-1}, J mol^{-1}K^{-1}
T	temperature	K
U	internal energy	J, J mol^{-1}
V	volume	m^3
W	work done	J, m^3 atm
W_i	weight percent of component i in a solution	–
Z	valency, coordination number	–

Greek Symbols

α	coefficient of volumetric thermal expansion	K^{-1}
α_i	Darken's α-function for component i in a solution	–
β	isothermal compressibility of a material	atm^{-1}
γ	ratio of C_P/C_V	–
γ_i	activity coefficient of component i in a solution	–
γ_i^0	Henry's Law constant for solute i in a binary solution	–
Γ_i	surface excess of component i	mol m^{-2}

δ	symbol for partial differential (e.g. δq, δW)	–
Δ	symbol indicating finite change of a state property (e.g. ΔG, ΔH, ΔS)	–
ε	single electrode potential	volt
ε_i	energy of a particle at ith level	J
ε_i^j	interaction coefficient giving influence of solute j on γ_i in a solution	–
η	efficiency of an engine	–
μ_i	chemical potential of component i in a solution	$J\ mol^{-1}$
ν	oscillation frequency of atoms in a crystal	s^{-1}
ρ	density of a material	$g\ cm^{-3}$, $kg\ m^{-3}$
σ	surface tension	$N\ m^{-1}$
Σ	summation sign	–
ϕ	electric potential	volt
Ω	a parameter	$J\ mol^{-1}$

Superscripts

0	denotes values at standard state for extensive state properties (e.g. S^0, G^0, H^0)
m	indicates change of value of a quantity due to mixing (e.g. $\bar{G}_i^m = \bar{G}_i - G_i^0$, $\Delta G^m = G - G^0$)
XS	value of a state property in excess of that in ideal solution (e.g. $\bar{G}_i^{XS} = \bar{G}_i - G_i^{id}$)

Subscripts

i, j, ...	for component i, j, ...
m	for melting
f	for formation of a compound from elements
T	at temperature T
Tr	for phase transformation
v	for vapourization

Others

| [ ] | metallic phase |
| ( ) | nonmetallic condensed phase (e.g. oxide, etc.) |

Chapter *1*

Introduction

1.1 Historical Perspective

Thermodynamics is a major discipline in science with widespread application in all branches of science and engineering. The subject was developed in the 19th and early 20th Century through the efforts of engineers, physicists and chemists. The laws of thermodynamics are general, but their applications are specific to various disciplines. Hence the emphasis, content and approach differ considerably from one branch to another in science and engineering disciplines.

Thermodynamics may be broadly classified into three: classical, statistical, and irreversible.

1. Classical (i.e. macroscopic or phenomenological) thermodynamics

Classical thermodynamics consists of the First, Second and Third Laws of thermodynamics. It was developed in the 19th Century and in the first decade of the 20th Century. The Zeroth Law is a later addition not recognised in many thermodynamics texts. Classical thermodynamics treats a substance as a continuum, ignoring behaviour of atoms and molecules.

2. Statistical thermodynamics

Statistical thermodynamics originated from the Kinetic Theory of Gases, which related pressure to average kinetic energy of molecules in an ideal gas. The application of probability theory, quantum theory and statistical mechanics allowed it to arrive at macroscopic thermodynamic relations from atomistic (i.e. microscopic) point of view. It was founded by Maxwell, Gibbs and Boltzmann in the late 19th Century, and was developed further in the 1920s and 1930s.

3. Irreversible thermodynamics

Irreversible thermodynamics deals with the application of thermodynamics to irreversible processes, and was first proposed by I. Prigogine in 1942.

Another way of classifying thermodynamics is:

1. Thermodynamics of *nonreactive* systems.
2. Thermodynamics of *reactive* systems, also known as chemical thermodynamics. It is often considered as a part of physical chemistry, which includes subjects such as reaction kinetics and diffusion besides chemical thermodynamics.

Since it is only the classical thermodynamics which has widespread application in Metallurgical Engineering and Materials Science and Engineering, we shall confine ourselves mostly to this, except for one chapter on statistical thermodynamics. For this reason, the term *thermodynamics* generally means classical thermodynamics. Thus, in Section 1.2, we will briefly review the historical development of classical thermodynamics.

1.2 History of Classical Thermodynamics

1.2.1 General

The literal meaning of *thermodynamics* is that it is a subject dealing with the relation between heat and motion. The invention of the steam engine in the 18th Century triggered the issue of the relation between heat and work amongst engineers and physicists. In 1798, Count Rumford observed that the heat generated during boring of cannons was approximately proportional to the work done. In 1840, Joules' classic and precise experiments provided confirmation of the above and led to the concept of *Mechanical Equivalent of Heat*. It was later generalized into *Law of Conservation of Energy*, which states that *"energy can neither be created nor destroyed; it can be only transformed from one form into another"*. These developments eventually led to the generalized definition of thermodynamics as a subject dealing with energy and its transformation from one form to another. This constituted the basis for the *First Law of Thermodynamics*.

Although mechanical energy can be completely converted into heat, it was observed that the reverse is not true, i.e. heat cannot be completely converted into work in a straightforward fashion. Quantitative explanation came first from the famous *Carnot's cycle*, derived by S. Carnot in 1824. These led to the enunciation of the *Second Law of Thermodynamics*, employing the concept of reversible and irreversible processes. Later in the mid-19th Century, Clausius and J.J. Thomson (Lord Kelvin) proposed *entropy* as the thermodynamic parameter for quantitative application of the Second Law.

In 1906, W. Nernst proposed his *Heat Theorem* on the basis of experimental evidence. It was later generalized by Max Planck as *"the entropy of any homogeneous substance (which is at complete internal equilibrium), may be taken as zero at 0 K"*. This is known as the *Third Law of Thermodynamics*.

1.2.2 Chemical Thermodynamics

From experimental evidence, in the late 18th Century, Lavoisier concluded that matter is conserved even when chemical reaction occurs. Thus the *Law of Conservation of Mass* came into existence. The 19th Century further witnessed major developments in chemistry, such as

- Dalton's atomic theory and reaction stoichiometry
- Avogadro's hypothesis and concept of g-mole
- The concept of chemical equilibrium
- The heats of reaction
- Faraday's Laws of Electrolysis

These provided the foundation for Physical Chemistry and *Chemical Thermodynamics*. In the late 19th Century, J.W. Gibbs proposed his famous free energy function, known as *Gibbs free energy*. The systematic application of this to thermodynamics by Gibbs and others firmly established the foundation of chemical thermodynamics. The subject attained maturity by early 20th Century.

Chemical thermodynamics is also based on the three laws of thermodynamics. However, it employs many other auxilliary relations also, mostly derived from Gibbs free energy considerations as well as other physico-chemical relations.

1.2.3 Metallurgical Thermodynamics

Broadly speaking, the application of primarily chemical thermodynamics to metals and materials has led to the development and growth of *Metallurgical Thermodynamics* or its later generalization as *Thermodynamics of Materials*.

The application of laws and relationships of thermodynamics requires experimental data such as enthalpies and free energies of reactions and processes, activity versus composition relationships in solutions, phase equilibria and phase diagrams. Processing of metals and ceramics is carried out primarily at high temperatures. Experimental measurements at high temperatures pose certain problems, not encountered at room temperature. Overcoming these require other technological developments, e.g. availability of high temperature materials, techniques, apparatus as well as instruments, many of which were not available earlier. Therefore, experimental measurements and assessment of thermodynamic data in high temperature metallurgical and materials area required several decades of efforts by a large number of scientists. This is the reason why the subject of Metallurgical Thermodynamics attained maturity as late as 1960/1970, although chemical thermodynamics was well established by 1930.

In the early days, several physical chemists took interest in metals and inorganic compounds as well as their behaviour at high temperatures. It is they who did pioneering work in thermodynamics of metals and materials.

1.3 Introductory Concepts and Definitions

As already stated, thermodynamics may be classified into: classical (i.e. macroscopic), statistical, and irreversible. However, it is classical thermodynamics that has general and large-scale applications. Therefore, this text will be concerned almost exclusively with this. The following introductory concepts and definitions are the starting points for classical thermodynamics.

1.3.1 System and Surrounding

Any portion of the Universe selected for consideration is known as the *system*. The rest of the Universe is its *surrounding*. Depending on the nature of interaction, systems have been classified into *open, closed and isolated*. An open system exchanges both matter and energy with its surrounding. A closed system exchanges only energy with its surrounding. An isolated system does not exchange either matter or energy with its surrounding. Hence the mass is fixed for a closed or isolated system during occurrence of a process. Laws and equations of thermodynamics are taught and comprehended more easily for the above. Hence, the readers are advised to visualize and keep in mind a closed system unless otherwise mentioned.

Classical thermodynamics deals with a system of *macroscopic size* (say, containing at least a million molecules), which we treat as a continuum, and ignore the details of structure. Because of these, the laws and equations of classical thermodynamics are independent of structure (i.e. structure insensitive). This is where its power and usefulness lies. However, the generalizations are valid under certain assumptions. In addition, some definitions have been proposed. It is necessary to understand these properly.

1.3.2 State of a System

By 'state' we shall mean the *macrostate* of a system in classical thermodynamics since this is what we shall be concerned with. The state of a system at any instant is defined by specifying all its *state variables* (which also include state properties), such as temperature, pressure, volume, and density. Thermodynamics deals with energy and its transformation (i.e. conversion) from one form to another. Simplifications are achieved by ignoring forms of energy which do not contribute significantly to the process under consideration. For example, in the absence of a significantly strong magnetic field of force, magnetic energy can be ignored. Surface energy is ignored if the specific surface area is small and, hence, surface energy term is not significant.

Pressure P, volume V, and temperature T are the most common state variables. The laws of thermodynamics have been proposed and expounded on the basis of the above. The resulting relations were then amended to include other variables as necessary.

For a substance of fixed mass, P, V and T are interrelated. For example, for an ideal gas,

$$PV = nRT \tag{1.1}$$

i.e.

$$v = \frac{V}{n} = \frac{RT}{P} \tag{1.2}$$

where n is number of moles (i.e. g-moles), R the universal gas constant, T the absolute temperature, and v the *molar volume*.

Now we have to distinguish between property and variable. The molar volume is a property of the substance. Of course, it can be varied. In that sense, it is a variable too. But pressure and temperature are not properties. They are variables, generally imposed by the surrounding. Therefore, variables may or may not be properties. It may also be noted here that state properties are also known as *state functions*.

The volume of a substance is proportional to its mass (or number of moles). On the other hand, P and T are independent of the mass of the system, and are known as *intensive variables*. Volume V is an *extensive property*. It can be made intensive by dividing it by mass or number of moles. Hence, the *molar volume (v) is an intensive property*. Density is an intensive property.

It is obvious that, if an equation contains both extensive property and intensive variable, then there must be a term denoting mass or number of moles. If the latter are missing, then there is an implicit assumption that the extensive property is for a fixed mass (say, a closed system). The general convention in chemical thermodynamics is to go for *molar properties*, which are intensive. This way, the relationships and functions become independent of the quantity of matter, and hence of more general applicability [e.g. Eq. (1.2)].

1.3.3 Thermodynamic Equilibrium

When we wish to characterize a state by specifying the values of the important intensive variables, it is assumed in thermodynamics that the system has come into equilibrium with respect to those variables. Then and only then the state can be properly defined in terms of these variables. A substance which is at thermodynamic equilibrium is known as a *thermodynamic substance*.

For a thermodynamic substance of fixed mass, the *equation of state* is

$$V = f(P, T) \tag{1.3}$$

Equation (1.2) is a specific case of Eq. (1.3) for ideal gas only. Another, and more general form of Eq. (1.3), is

$$v = \frac{V}{n} = f(P, T) \tag{1.4}$$

It should be emphasized here that Eqs. (1.3) and (1.4) are applicable to a substance of constant composition and structure, i.e. its nature remains the same during change of pressure and temperature. For every substance, except for ideal gas, the function is different.

When the system is at equilibrium with respect to an intensive variable, the magnitude of the variable throughout the system is the same (i.e. uniform). The term *equilibrium* has originated from the Latin word "acquilibrium", meaning equal weight (e.g. as in a weighing balance). In general, it refers to a state, which is not changing with time.

The mechanical interaction of the system with the surrounding is represented by pressure, in the absence of a field of force (electric, magnetic etc.). Here, *mechanical equilibrium* means pressure equilibrium, i.e. uniformity of pressure throughout the system. Thus, there is no tendency for the pressure to change with time anywhere in the system.

Similarly, at *thermal equilibrium*, temperature should be uniform throughout the system. Only then there will be no tendency for heat to flow from one part of the system to another, i.e. no thermal change takes place in the system.

For purely physical processes, such as expansion/compression of a substance, thermodynamic equilibrium consists of mechanical and thermal equilibria. But for physico-chemical processes and chemically reactive systems, thermodynamic equilibrium would also require attainment of *physico-chemical/chemical equilibrium* in addition to mechanical and thermal equilibria. This means attainment of uniform *chemical potential* besides uniformity of pressure and temperature in the system. Detailed discussions of chemical potential will be taken up in Chapter 10.

The system is at its most stable state when it is at thermodynamic equilibrium. It will stay that way unless disturbed. An open or a closed system interacts with its surrounding. There, the equilibrium is always with respect to the surrounding. If the surrounding changes, the state of equilibrium of the system will also change.

Sometimes a system is not at complete thermodynamic equilibrium, but at partial or pseudo equilibrium. Thermodynamics is capable of handling such cases. One example is *metastable state*, where the system undergoes change so slowly that there will be no perceptible change in a long period of time (say, several years/centuries/millennia). For instance, if a mixture of hydrogen and oxygen gas is kept at room temperature, very little detectable reaction would occur even in years, although it is not at chemical equilibrium.

Cementite (Fe_3C) in iron-carbon system is metastable, but can stay for perhaps several centuries or millennia at room temperature. In both the above cases, raising the temperature increases the rate enormously, and allows attainment of chemical equilibria in a short period of time.

Another example is partial chemical equilibria in some complex reactive systems, where some reactions attain equilibrium whereas some others do not.

1.3.4 Characterization of Systems—Some elucidations

We have broadly classified system—as open, closed and isolated already. It has also been stated that processes may be purely physical or physico-chemical/chemical. In the former case, we are dealing with a nonreactive system, whereas, when physico-chemical processes and/or chemical reactions occur, the system is a reactive one. Some further comments are now noted to elaborate the above observations.

A system consisting of one *phase* only (say, water) is a *single-phase system*, which is known as a *homogeneous system*. At 0°C, ice and water can co-exist at equilibrium. Then it is a *heterogeneous system*. However, the above classifications do not provide full description of a system. Further classifications are required for application of the laws of thermodynamics.

A substance in a closed or isolated system not only has fixed mass, but has fixed overall composition as well. Consider water in a closed vessel. On cooling below 0°C, it is transformed into ice. Again, on heating above 100°C, it is transformed into a gas. Therefore, the state of aggregation of H_2O (broadly speaking, structure of H_2O) depends on temperature. Similarly, above 910°C, the structure of pure iron changes from α-Fe (i.e. BCC) to γ-Fe (i.e. FCC). These changes are known as *phase changes*, and are physico-chemical processes. The system has *one-component* only (H_2O or Fe), and is also referred to as *Unary system*.

A *multicomponent system* obviously has more than one component. An aqueous solution of NaCl, for example, has two components, (viz. NaCl and H_2O), and is known as a *binary* system. Similarly, we have *ternary, quarternary* etc., all coming under multicomponent system. If the NaCl solution is cooled below 0°C, pure ice crystals would start forming. As temperature is lowered further, more and more ice forms. This increases concentration of NaCl in residual water. Hence, in this two-phase system, composition of water is changing. However, composition of the entire system in terms of number of moles of H_2O and NaCl does not change, if it is a closed system.

In contrast, in an open system, both mass and composition of the system can change. For example, if we consider only the aqueous NaCl solution as our system (even in a closed vessel) in the two-phase ice + aqueous solution situation, then both mass and composition of the solution are changing with change of temperature. If pure water is heated in an open beaker, its mass will keep decreasing with vaporization. If it is an aqueous solution, then besides mass, concentration of NaCl in solution will keep increasing with vaporization.

If water is heated to a very high temperature, the steam dissociates into hydrogen and oxygen, and we have to be concerned with this chemical reaction in the gaseous state. Then, if the system is closed, its total mass will remain constant. Not only that, its composition in terms of total number of moles of hydrogen and oxygen (in elemental form and combined as H_2O) will remain constant. Therefore, the meaning of the term *composition* has to be interpreted that way.

To sum up, in classical thermodynamics, a system can be classified as:

1. Open, closed or isolated
2. Unary or multicomponent
3. Reactive or nonreactive
4. Homogeneous or heterogeneous

These are required for application of thermodynamics. Some of these will again be elucidated in subsequent chapters. However, it is hoped that these introductory comments would be useful to readers.

1.4 Values of Some Physical Constants

Absolute temperature of ice point (0°C) = 273.15 K

$$1 \text{ calorie} = 4.184 \text{ joules}$$

Avogadro's number (N_0) = 6.023×10^{23} per mole

Boltzmann's constant (k_B) = 1.3805×10^{-23} joules/K/molecule

Faraday's constant (F) = 96,500 coulombs/g-equivalent (i.e. joules/volt/g-equivalent)

1 Standard atmosphere = 760 mm Hg
$$= 1.013 \text{ bar}$$
$$= 1.013 \times 10^5 \text{ pascal}$$

Universal Gas Constant (R) = 8.314 joules/mol/K
$$= 82.06 \text{ cm}^3\text{-atm/mol/K}$$
$$= 82.06 \times 10^{-6} \text{ m}^3\text{-atm/mol/K}$$

1.5 Summary

1. Thermodynamics may be broadly classified as: classical (i.e. macroscopic or phenomenological), statistical, and irreversible.

2. Most applications are based on classical thermodynamics. Hence, the present text is primarily concerned with it. Also, generally, the term *thermodynamics* denotes classical thermodynamics.

3. The three laws of thermodynamics constitute the foundations of thermodynamics.

4. Thermodynamics deals with energy and its transformation from one form to another.

5. Metallurgical Thermodynamics, or its later generalization, Thermodynamics of Materials, belongs to chemical thermodynamics, which deals with reactive systems.

6. A system is classified as: open, closed and isolated. The rest of the Universe is its surrounding.

7. State variables are extensive or intensive.

8. Molar properties are intensive properties (or variables).

9. Pressure, temperature and volume are common state variables.

10. A state can be defined by state variables only at thermodynamic equilibrium. It then constitutes a thermodynamic substance.

11. A system needs characterization before its thermodynamic treatment.

Chapter 2

First Law of Thermodynamics

Two important thermodynamic quantities have been proposed on the basis of the First Law of Thermodynamics, viz. *Internal Energy U* and *Enthalpy H*. Both are energy terms and are state properties.

2.1 Internal Energy

Imagine a system. Suppose a quantity of *heat q* is supplied to it from the surrounding, and as a consequence the system does an amount of *work W* on the surrounding. If these are the only two energy terms involved in the process, then *the Law of Conservation of Energy* demands that q be equal to W. However, experiments have revealed that, commonly,

$$q - W \neq 0$$

In order to save the Law of Conservation of Energy, it was proposed that any system possess an internal energy U, which is a hidden form of energy stored in it. The difference between q and W was attributed to change in U (i.e. ΔU) during the process, i.e.

$$\Delta U = q - W \tag{2.1}$$

It should be noted that according to mechanics, it is actually a force and not a system that does work. The concept that a system does work arose from the historical context of development of classical thermodynamics, viz. conversion of heat into mechanical work by a heat engine. Also, it should be further noted that in school text books and modern thermodynamics texts in some areas, W is taken as positive, if the work is done on the system, which makes $\Delta U = q + W$. However, in traditional thermodynamics, physical chemistry and chemical thermodynamics texts, the convention, as in Eq. (2.1), is followed, and, therefore, we are retaining the same.

Equation (2.1) is the mathematical statement of the *First Law of Thermodynamics*. On the basis of our present day knowledge, the internal energy consists of

1. macroscopic kinetic energy due to motion of the system as a whole,

2. potential energy of the system due to its position in a force field,
3. kinetic energies of atoms and molecules in the form of translation, rotation and vibration (i.e. thermal energy),
4. energies of interaction amongst atoms and molecules (a form of potential energy),
5. coulombic energy of interaction amongst electrons and nucleii in atoms, and
6. energy contents of electrons and nucleii of atoms.

In conventional chemical thermodynamics which we shall be concerned with, only kinetic energies of atoms and molecules, and interactions amongst atoms and molecules, i.e. items (3) and (4) above, are considered to be important since changes occurring in them principally contribute to ΔU. However, changes in other forms of internal energy also assume importance in non-conventional processes. For example, (5) and (6) would be significant in nuclear reactions.

It is also to be understood that the absolute value of U for a system is not known. All we can determine is ΔU. It may also be noted that U of a substance (i.e. material) depends on its mass, *nature (i.e. chemical composition* and *structure)* as well as *temperature.* For a material of fixed mass, composition and structure, U is a function of temperature only.

2.2 Statement of the First Law

Equation (2.1) is the mathematical statement of the first law for a finite process. For an *infinitesimal (i.e. differential) process*, the statement is

$$dU = \delta q - \delta W \tag{2.2}$$

2.2.1 Significance of the First Law

1. It is based on the Law of Conservation of Energy.
2. It brought in the concept of *internal energy.*
3. It separates heat interaction and work interaction between the system and the surrounding as two different terms.
4. It treats internal energy as a state property, i.e. dU in Eq. (2.2) is an exact differential. Hence, ΔU or dU would depend only on the initial and final states of the system, and not on the path by which the system has moved from one state to another. In contrast, q and W are energy in transition, and therefore, are not state properties/ variables but are dependent on the path. Since the path may be known or unknown, and is likely to depend on several variables (known and unknown), infinitesimal changes in q and W in Eq. (2.2) are designated by δq and δW, not dq and dW. The readers are advised to exercise particular care to maintain this distinction.

2.2.2 Proof of the First Law

The concept of internal energy was proposed assuming the validity of the Law of Conservation of Energy. ΔU is just the difference between q and W. There is no direct proof of existence of U as a state property. However, there are plenty of indirect experimental evidence of its existence and is a well-established quantity.

In the early days of thermodynamics in the 19th Century, when the First Law was proposed, the nature of U as a state property was inferred from experiments on *cyclic processes*, where the system was changed but ultimately brought back to the same initial state. It was found that for such cyclic processes,

$$\sum_{\text{cycle}} q - \sum_{\text{cycle}} W = 0 \qquad (2.3)$$

provided q and W are expressed in the same energy unit (say, joule). This means that $\sum_{\text{cycle}} \Delta U = 0$, i.e. U is a state property.

As discussed in Chapter 1, in conventional classical thermodynamics, we are generally concerned with P, V, T as state variables. In chemical thermodynamics, the chemical composition and structure of substance in the system also are variables. However, variation of thermodynamic quantities with composition and structure will be taken up later. For the time being, we shall assume the substance to retain its initial composition and structure all throughout the process (i.e. *a nonreactive system*).

In Chapter 1, we have further stated that P, V, T for a system of fixed mass are interdependent, with only two of them acting as independent variables. Hence, for a system of *fixed mass* (imagine a *closed system* for easy comprehension),

$$U = f(V, T) \qquad (2.4)$$

or

$$U = f(P, T) \qquad (2.5)$$

or

$$U = f(P, V) \qquad (2.6)$$

Invoking properties of exact differentials, from Eqs. (2.4)–(2.6), respectively,

$$dU = \left(\frac{\partial U}{\partial V}\right)_T dV + \left(\frac{\partial U}{\partial T}\right)_V dT \qquad (2.7)$$

or

$$dU = \left(\frac{\partial U}{\partial P}\right)_T dP + \left(\frac{\partial U}{\partial T}\right)_P dT \qquad (2.8)$$

or

$$dU = \left(\frac{\partial U}{\partial P}\right)_V dP + \left(\frac{\partial U}{\partial V}\right)_P dV \qquad (2.9)$$

2.3 Reversible Processes

In a reversible process, the system is displaced from equilibrium infinitesimally, and then allowed enough time to attain a new equilibrium; and again displaced infinitesimally, and thus the process continues. Thus, a reversible process may be defined as *the hypothetical passage of a system through a series of equilibrium stages.* A reversible process is very slow and mostly impracticable. No going process is reversible in a strict sense. The terminology *reversible process* arises from the fact that it can be reversed by an infinitesimal change of a variable in the opposite direction.

The concept of reversible process is at the heart of thermodynamics. It may be explained by the traditional *cylinder-piston analogy.* The gas inside the cylinder constitutes the system and everything else is the surrounding. One way to expand the gas is to decrease the external pressure. Imagine lowering of the external pressure from P_1 to P_2. During expansion, the temperature of gas goes down, causing heat flow into it from outside through the cylinder wall. Finally, a new equilibrium is attained when the gas pressure inside is P_2 and the temperature is again equal to that of the surrounding (T_{sur}).

The overall process is isothermal since both initial and final temperatures are equal to T_{sur}. However, during expansion, the gas temperature will be lower than T_{sur}. Figure 2.1 depicts the process as temperature vs. time plot. If pressure is lowered rapidly to $P - \Delta P$, where ΔP is a finite change in P, the process is irreversible and variation of temperature with time is unknown, although a hypothetical curve is drawn in Fig. 2.1, just for visualization.

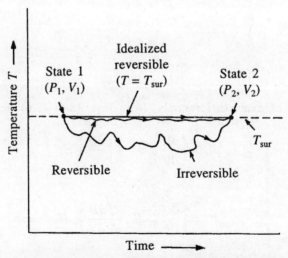

Fig. 2.1 Temperature vs. time curves for isothermal processes (schematic); path is hypothetical for irreversible process.

Process, on the other hand, occurs reversibly if the pressure is decreased a little to $P - dP$, where dP is an infinitesimally small change in pressure. Then sufficient time is allowed for the gas to attain T_{sur}, and again the pressure is decreased a little, and time allowed. Here we are carrying out the process slowly through a series of equilibrium stages. It should be noted that the temperature is always close to T_{sur} in the reversible process. The mathematically limiting situation for curve 2 is the constant temperature line at $T = T_{sur}$, which is the idealized version of an *isothermal reversible process*. Figure 2.2 illustrates these.

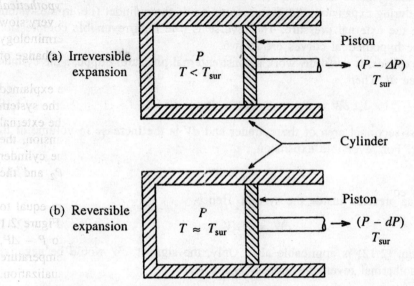

Fig. 2.2 Irreversible and reversible expansion of a gas at any instant of time.

Now, consider isothermal expansion-compression of n moles of an ideal gas. Figure 2.3 shows the process on a *P-V* diagram. The *equation of state* is

$$PV = nRT$$

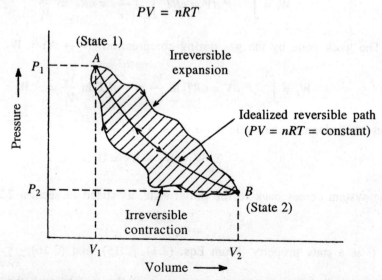

Fig. 2.3 Pressure-volume diagram for isothermal expansion-compression cycle of an ideal gas in a cylinder-piston system (schematic); *P* is inside pressure; path is hypothetical for irreversible cycle.

which is the same as Eq. (1.1), and the corresponding curve is for isothermal reversible process at temperature T. The irreversible expansion and compression cannot be depicted on the P-V diagram since their paths are undefined. Moreover, for irreversible process, the pressure of gas in the cylinder is not expected to be uniform. However, it is qualitatively correct to say that, during expansion, the gas pressure inside the cylinder (i.e. in the system) will be higher than the external pressure. The reverse is true for irreversible compression. On this basis, some hypothetical curves are shown.

When the gas expands, it performs work against external pressure. Suppose the cylinder moves by a distance dl. Then,

$$\delta W = P_{ext} \cdot A \cdot dl = P_{ext} \cdot dV \tag{2.10}$$

where A is the cross-sectional area of the cylinder and dV is the increase in volume of the gas in the cylinder. For reversible expansion,

$$P_{ext} \approx P_{int} = P \tag{2.11}$$

where P denotes gas pressure inside the system. Hence,

$$\delta W = P\,dV \tag{2.12}$$

For compression, Eq. (2.12) is applicable also. Only, the sign of δW would be negative. Consider the isothermal reversible process again. Here,

(i) The work done by the gas during expansion state A → B (as in Fig. 2.3) = W_1, where

$$W_1 = \int_{V_1}^{V_2} P\,dV = nRT \int_{V_1}^{V_2} \frac{dV}{V} = nRT \ln \frac{V_2}{V_1} \tag{2.13}$$

(ii) The work done by the gas during compression (B → A) = W_2, where

$$W_2 = \int_{V_2}^{V_1} P\,dV = nRT \ln \frac{V_1}{V_2} = -nRT \ln \frac{V_2}{V_1} = -W_1 \tag{2.14}$$

Thus, net work done in the cycle A → B → A is

$$\sum_{cycle} W = W_1 + W_2 = 0 \tag{2.15}$$

Since the system comes back to the initial state, as stated in section 2.2.2,

$$\sum_{cycle} \Delta U = 0 \tag{2.16}$$

because U is a state property. From Eqs. (2.1), (2.15), and (2.16), $\sum_{cycle} q = 0$.

This means that there is no net interaction of the system with the surrounding for a cyclic reversible process. *Hence, it does not leave any permanent change in system or surrounding during reversing.* This is an important conclusion.

It is obvious that for irreversible processes, the above is not true and some net heat and work interaction with the surrounding would occur for cycle A $\rightarrow$ B $\rightarrow$ A. The shaded area is the net work done during the irreversible cycle A $\rightarrow$ B $\rightarrow$ A. The presence of friction between the piston and the cylinder will introduce irreversibility as well. Therefore, for reversibility, friction should be absent.

It should be noted again that the path is not known for an irreversible process. Hence, it cannot be shown in a diagram. Therefore, the irreversible processes shown in Figs. 2.1 and 2.3 are hypothetical.

2.4 Calculation of *W* and *q* for Reversible Processes

Both *W* and *q* are dependent on path. Hence, they can be calculated for a process if the entire path is known, and this is possible only for reversible processes, as explained in Section 2.3. In general,

$$\delta W = P\,dV + \delta W' \tag{2.17}$$

where

δW = infinitesimal total work done by the system, as in Eq. (2.2)
$P\,dV$ = infinitesimal work done by the system against pressure
$\delta W'$ = infinitesimal quantity of other forms of work done by the system (electric, magnetic, etc.).

We shall consider $\delta W'$ when we discuss electrochemical cells or surfaces (in Chapters 13 and 14). Till then, in all our subsequent discussions, we shall assume $\delta W' = 0$. Hence, $\delta W = P\,dV$, and combining this with Eq. (2.2),

$$dU = \delta q - P\,dV, \text{ i.e. } \delta q = dU + P\,dV \tag{2.18}$$

For a finite reversible process from state 1 to state 2,

$$W = \int_{V_1}^{V_2} P\,dV \tag{2.19}$$

and would depend on the *P-V* relationship. Let us consider the following reversible process paths:

(i) At constant volume (i.e. *isochoric*), $W = 0$; hence from Eq. (2.1),

$$\Delta U = q \tag{2.20}$$

(ii) At constant pressure (i.e. *isobaric*),

$$W = P(V_2 - V_1) \tag{2.21}$$

(iii) At constant temperature (i.e. *isothermal*)

On the basis of experimental measurements by Joule and others in the 19th Century, it was proposed that U is a function of temperature only for an ideal gas. Later, kinetic theory of gases also predicted the same. Hence for the isothermal expansion-compression of an ideal gas,

$$\Delta U = 0, \text{ i.e. } W = q \qquad (2.22)$$

It should be kept in mind that Eq. (2.22) is applicable only to a fixed mass of gas since any change in mass will change internal energy even at a constant temperature.

(iv) Adiabatic:

$$q = 0, \text{ i.e. } \Delta U = - W \qquad (2.23)$$

2.5 Enthalpy

Enthalpy H is defined as

$$H = U + PV \qquad (2.24)$$

Hence, by definition, H is a state property, since U and V are state properties and P is a state variable. Differentiating Eq. (2.24), we get

$$dH = dU + P\,dV + V\,dP \qquad (2.25)$$

i.e.

$$dH = dU + P\,dV \text{ at constant } P \qquad (2.26)$$

If $\delta W' = 0$, then by combining Eq. (2.26) with Eq. (2.18), we obtain

$$dH = (\delta q - P\,dV) + P\,dV = \delta q \qquad (2.27)$$

Hence, for a finite process at constant pressure from state 1 to state 2,

$$q = \int_{\text{state 1}}^{\text{state 2}} dH = H_2 - H_1 = \Delta H \qquad (2.28)$$

Since most processes occur or are carried out under conditions where Eq. (2.28) is exactly or approximately valid, Eq. (2.28) constitutes the basis for *process heat balance*, which is very important for manufacturing industries as well as analysis of processes. Energy is costly, and should be accounted for and saved as much as possible. The bulk of the energy supply goes to meet process heat requirements. Without process heat balance, energy audit is not possible. Further, Eq. (2.28) allows experimental determination of ΔH of a process by measurement of q by a calorimeter. Since H is a state property, ΔH (i.e. $H_2 - H_1$) can be tabulated in thermodynamic data sources and used by all.

Many texts, especially the older ones, therefore, refer to enthalpy *as heat content of a substance* due to the above equality. *However, the latter terminology is not scientifically correct since heat is energy in transition and not a state property.*

As stated in Section 2.4, *internal energy of an ideal gas is a function of temperature only*. At constant temperature (i.e. $dT = 0$), therefore, $dU = 0$, and Eq. (2.8) gets simplified into

$$\left(\frac{\partial U}{\partial P}\right)_T = 0 \tag{2.29}$$

Differentiating Eq. (2.24) with respect to P, we obtain

$$\left(\frac{\partial H}{\partial P}\right)_T = \left(\frac{\partial U}{\partial P}\right)_T + \left[\frac{\partial(PV)}{\partial P}\right]_T \tag{2.30}$$

In ideal gas at constant T, $PV = $ constant; so, $d(PV) = 0$. From all these above,

$$\left(\frac{\partial H}{\partial P}\right)_T = \left(\frac{\partial U}{\partial P}\right)_T = 0 \tag{2.31}$$

In other words, *enthalpy of an ideal gas is independent of pressure at constant temperature. Similarly, enthalpy is independent of volume. Hence, H is a function of T only for a fixed mass of a substance.* This is of great help in calculating the process heat balance since in most metallurgical processes, the gas phase is essentially ideal.

EXAMPLE 2.1 The Equation of State of a gas is given by the expression

$$\left(P + \frac{a}{V^2}\right)(V - b) = RT \tag{E.2.1}$$

where a, b are constants. Prove that

$$\left(\frac{\delta T}{\delta P}\right)_V \left(\frac{\delta P}{\delta V}\right)_T \left(\frac{\delta V}{\delta T}\right)_P = -1$$

Solution The partial differentiation of Eq. (E.2.1) yields

$$\left(\frac{\delta T}{\delta P}\right)_V = \frac{V - b}{R} \tag{E.2.2}$$

$$\left(\frac{\delta P}{\delta V}\right)_T = \frac{1}{V - b}\left[\frac{2a(V - b)}{V^3} - \left(P + \frac{a}{V^2}\right)\right] \tag{E.2.3}$$

$$\left(\frac{\delta V}{\delta T}\right)_P = \frac{R}{\left(P + \dfrac{a}{V^2}\right) - \dfrac{2a(V - b)}{V^3}} \tag{E.2.4}$$

Combining Eqs. (E.2.2)–(E.2.4), we obtain

$$\left(\frac{\delta T}{\delta P}\right)_V \left(\frac{\delta P}{\delta V}\right)_T \left(\frac{\delta V}{\delta T}\right)_P = -1$$

2.6 Heat Capacity

It is the quantity of heat required to raise the temperature of a substance by 1°C (i.e. 1 K). In thermodynamics, the *molar heat capacity* (C), i.e. heat capacity of 1 g-mole of a substance is most widely employed. Thus,

$$C = \frac{\delta q}{dT} \tag{2.32}$$

δq is heat required to raise the temperature of 1 mole by dT. Since δq depends on path, C is a meaningless quantity unless the path is specified. The following have proved to be useful.

The molar heat capacity *at constant volume* is given by

$$C_V = \left(\frac{\delta q}{\partial T}\right)_V = \left(\frac{\partial U}{\partial T}\right)_V \tag{2.33}$$

since at a constant volume, $\delta q = dU$. Similarly, molar heat capacity *at constant pressure* (C_P) is given by

$$C_P = \left(\frac{\partial q}{\partial T}\right)_P = \left(\frac{\partial H}{\partial T}\right)_P \tag{2.34}$$

since at constant P, $\delta q = dH$. In Eqs. (2.33) and (2.34), U and H are molar values of internal energy, and enthalpy, respectively.

From Eq. (2.34),

$$dH = C_P \, dT \tag{2.35}$$

Integrating Eq. (2.35) between temperatures T_1 and T_2, we get

$$H_{T_2} - H_{T_1} = \int_{T_1}^{T_2} C_P \, dT \tag{2.36}$$

It should be noted that U and H are molar quantities here. Earlier, they referred to internal energy and enthalpy of the entire system. Both conventions will be followed. So, U and H may be molar values or for the entire system, depending on what is being discussed.

$C_P > C_V$ since C_P includes heat required to do work against pressure also, besides raising temperature. From Eqs. (2.33) and (2.34),

$$C_P - C_V = \left(\frac{\partial H}{\partial T}\right)_P - \left(\frac{\partial U}{\partial T}\right)_V = \left[\frac{\partial(U+PV)}{\partial T}\right]_P - \left(\frac{\partial U}{\partial T}\right)_V$$

$$= \left(\frac{\partial U}{\partial T}\right)_P - \left(\frac{\partial U}{\partial T}\right)_V + P\left(\frac{\partial V}{\partial T}\right)_P \tag{2.37}$$

Again, differentiating Eq. (2.7) with respect to T at constant P, we get

$$\left(\frac{\partial U}{\partial T}\right)_P = \left(\frac{\partial U}{\partial V}\right)_T \left(\frac{\partial V}{\partial T}\right)_P + \left(\frac{\partial U}{\partial T}\right)_V \tag{2.38}$$

Combining Eqs. (2.37) and (2.38), we get

$$C_p - C_V = \left(\frac{\partial U}{\partial V}\right)_T \left(\frac{\partial V}{\partial T}\right)_P + P\left(\frac{\partial V}{\partial T}\right)_P = \left(\frac{\partial V}{\partial T}\right)_P \left[\left(\frac{\partial U}{\partial V}\right)_T + P\right] \tag{2.39}$$

2.7 Ideal Gases

2.7.1 Heat Capacity Relations

For 1 mole of an ideal gas, from Eq. (1.1),

$$\left(\frac{\partial V}{\partial T}\right)_P = \frac{R}{P} \tag{2.40}$$

Also, from Section 2.5,

$$\left(\frac{\partial U}{\partial V}\right)_T = 0 \tag{2.41}$$

Combining Eqs. (2.39)–(2.41), therefore, we get

$$C_P - C_V = R \tag{2.42}$$

2.7.2 Adiabatic Expansion and Compression

For an adiabatic process, $\delta q = 0$. Hence,

$$dU = - \delta W = - P\,dV \tag{2.43}$$

Since $PV = nRT$ for n moles of ideal gas,

$$dU = -\frac{nRT}{V}\, dV \tag{2.44}$$

Again, from Eq. (2.33),

$$dU = nC_V\, dT \tag{2.45}$$

Combining Eqs. (2.44) and (2.45), and integrating between $T = T_1$, $V = V_1$, and $T = T_2$, $V = V_2$, we get

$$nC_V \ln\left(\frac{T_2}{T_1}\right) = nR \ln\left(\frac{V_1}{V_2}\right)$$

i.e.

$$\frac{T_2}{T_1} = \left(\frac{V_1}{V_2}\right)^{R/C_V} \tag{2.46}$$

Let $C_P/C_V = \gamma$. Then, from Eq. (2.42),

$$\frac{R}{C_V} = \gamma - 1 \tag{2.47}$$

$$\frac{T_2}{T_1} = \left(\frac{V_1}{V_2}\right)^{\gamma-1} \tag{2.48}$$

From Gas' Laws,

$$\frac{T_2}{T_1} = \frac{P_2 V_2}{P_1 V_1} \tag{2.49}$$

Combining Eqs. (2.48) and (2.49), we obtain

$$\frac{P_2}{P_1} = \left(\frac{V_1}{V_2}\right)^{\gamma}, \text{ i.e. } PV^{\gamma} = m = \text{a constant} \tag{2.50}$$

Figure 2.4 compares P-V relationships for isothermal and adiabatic *reversible expansion* of an ideal gas on the P-V diagram, starting from the same initial state $A(P_1, V_1, T_1)$ to the same final pressure P_2. The final volume is larger for the isothermal process.

2.7.3 Work Done in Reversible Adiabatic Expansion

For the reversible adiabatic expansion (Fig. 2.4),

$$W = \int_{V_1}^{V_2} P\, dV = m \int_{V_1}^{V_2} \frac{dV}{V^{\gamma}} \tag{2.51}$$

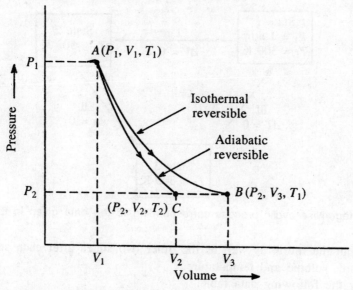

Fig. 2.4 *P-V* diagrams for isothermal and adiabatic reversible expansion of an ideal gas from the same initial state (schematic).

$$= m\left(\frac{V_2^{1-\gamma} - V_1^{1-\gamma}}{1 - \gamma}\right) \qquad (2.52)$$

Noting from Eq. (2.50) that

$$m = P_1 V_1^{\gamma} = P_2 V_2^{\gamma} \qquad (2.53)$$

we obtain

$$W = \frac{P_2 V_2 - P_1 V_1}{1 - \gamma} \qquad (2.54)$$

From Eq. (2.45),

$$\Delta U = U_2 - U_1 = nC_V(T_2 - T_1) \qquad (2.55)$$

Also, $\Delta U = -W$ for adiabatic process (Eq. 2.23).

Sometimes, an equation of the type:

$$PV^n = \text{a constant} \qquad (2.56)$$

is employed. Here, n is not equal to γ. It is known as a *polytropic process*.

EXAMPLE 2.2 Consider a three-step cyclic process, as shown in Fig. 2.5. The working substance is 1 kg-mol of a diatomic ideal gas initially at 300 K and 1 atm. The gas is heated at constant volume to 500 K, then expanded adiabatically to the initial temperature of 300 K, and finally compressed isothermally to initial pressure of 1 atm. Each step occurs reversibly.

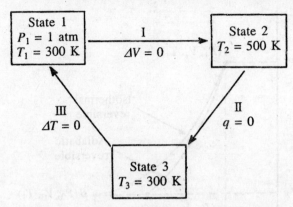

Fig. 2.5 Stage-wise cyclic process corresponding to the data given in Example 2.2.

(a) Find out the initial as well as the other two states after each step, in terms of pressure, volume and temperature.

(b) Verify the following data table.

Path	ΔU	ΔH	W	q
	(all values in kilojoules)			
I	4157	5820	0	4157
II	−4157	−5820	4157	0
III	0	0	−3185	−3185
Cycle (I + II + III)	0	0	972	972

Solution (a) The P, V, T of various states:

State 1: $P_1 = 1$ atm, $T_1 = 300$ K

$$V_1 = 10^3 \times (22.4 \times 10^{-3}) \times \frac{300}{273} \text{ m}^3 = 24.62 \text{ m}^3$$

State 2: $T_2 = 500$ K, $V_2 = 24.62$ m³, $P_2 = 1 \times \dfrac{500}{300} = 1.6$ atm

State 3: $T_3 = 300$ K

From Eq. (2.48),

$$\frac{T_2}{T_3} = \left(\frac{V_3}{V_2}\right)^{\gamma-1}$$

For a diatomic gas, $\gamma = \dfrac{7}{5} = 1.4$

Hence,

$$V_3 = V_2 \left(\frac{T_2}{T_3}\right)^{1/(\gamma-1)} = 24.62 \left(\frac{500}{300}\right)^{1/(1.4-1)} = 88.29 \text{ m}^3$$

$P_3V_3 = nRT_3$, where n = No. of moles = 10^3

Thus,

$$P_3 = \frac{10^3 \times (82.06 \times 10^{-6}) \times 300}{88.29} = 0.279 \text{ atm}$$

[*Note:* $R = 82.06 \times 10^{-6}$ m^{3}atm/mol.]

(b) *Step I:* $\Delta U = q - W = q - 0$ (isochoric) = q

$$= nC_V (T_2 - T_1) = 10^3 \times (2.5R) \times (500 - 300) \times 10^{-3}$$

$$= 10^3 \times 2.5 \times 8.314 \times 200 \times 10^{-3} = 4157 \text{ kJ}$$

[*Note:* $R = 8.314$ J/mol.]

$$\Delta H = nC_P(T_2 - T_1) = 10^3 \times 3.5 \times 8.314 \times 200 \times 10^{-3} = 5820 \text{ kJ}$$

Step II: $\Delta U = q - W = 0$ (adiabatic) $- W = -W$

$$= nC_V(T_3 - T_2) = 10^3 \times (2.5 \, R) \, (300 - 500) \times 10^{-3}$$

$$= -4157 \text{ kJ}$$

Therefore, $W = 4157$ kJ, $\Delta H = -5820$ kJ

Step III: $\Delta U = q - W = 0$ (isothermal)

Thus,

$$q = W = \int_{V_3}^{V_1} P \, dV = nRT \ln \frac{V_1}{V_3} \quad \text{(from Eq. (2.13))}$$

$$= 10^3 \times 8.314 \times 300 \left(\ln \frac{24.62}{88.29}\right) \times 10^{-3}$$

$$= -3185 \text{ kJ}$$

From Eq. (2.24), $H = U + PV$. Hence,

$$\Delta H = \Delta U + \Delta(PV)$$

For isothermal process and an ideal gas, $\Delta U = 0$, and $\Delta(PV) = \Delta(nRT) = 0$. Thus, $\Delta H = 0$.

2.8 Summary

1. The statements of the First Law of Thermodynamics are:

$$\Delta U = q - W, \quad \text{for a finite process}$$
$$dU = \delta q - \delta W, \quad \text{for an infinitesimal process.}$$

2. U is internal energy of the system, and is a state property. U is a function of temperature only for a fixed mass of an ideal gas.

3. q is heat absorbed by the system from the surrounding, and W is work done by the system on the surrounding.

4. q and W are energy in transition, are path dependent and, not state properties/state variables.

5. $\delta W = P\,dV + \delta W'$, where $P\,dV$ is work done against pressure, and $\delta W'$ is any other form of work (electric, magnetic, surface etc.). For all topics up to Chapter 12, $\delta W' = 0$, and $\delta W = P\,dV$.

6. For a reversible process, the path is known. Hence, W and q can be calculated. For a finite reversible process going from state 1 to state 2, if $\delta W' = 0$, then

$$W = \int_{V_1}^{V_2} P\,dV, \quad \text{where } P \text{ is the pressure in the system.}$$

7. Enthalpy H is defined as $H = U + PV$, and is a state property. At constant pressure, $\delta q = dH$, i.e. $q = \Delta H$.

8. For an ideal gas of fixed mass, H is a function of temperature only.

9. (a) Molar heat capacity at constant volume $= C_V = \left(\dfrac{\delta U}{\delta T}\right)_V$

   (b) Molar heat capacity at constant pressure $= C_P = \left(\dfrac{\delta H}{\delta T}\right)_P$

10. For an ideal gas, $C_P - C_V = R$.

11. For adiabatic expansion-compression of an ideal gas, $PV^\gamma = $ a constant, where $\gamma = C_P/C_V$.

PROBLEMS

2.1 Calculate the difference between ΔH and ΔU per mole (i.e. g-mole) of the reaction:

$$2H_2\,(g) + O_2\,(g) = 2H_2O\,(g)$$

at 500°C and 1 atm pressure.

2.2 One mole of an ideal monatomic gas is compressed from an initial volume of 20 litres to 10 litres by a reversible polytropic process, following the law:

$PV^{1.2}$ = constant. The initial temperature of the gas was 100°C.

Calculate:

(a) W, q for the process

(b) ΔU, ΔH between the final and initial states of the gas.

2.3 3.5 moles of an ideal diatomic gas is initially at 500 K and 18 atm pressure. It undergoes reversible isothermal expansion from 18 to 8 atm pressure, and then further reversible adiabatic expansion from 8 atm to 1.6 atm. Calculate: W, q, ΔU, ΔH for each stage.

2.4 Calculate W, q, ΔU, ΔH for reversible expansion of 2 moles of oxygen gas by the adiabatic and isobaric paths. Assume oxygen to behave as ideal gas. At the initial state, $P = 2$ atm, $T = 273$ K. The final volume is double that of the initial volume.

2.5 A system comprises one mole of an ideal monatomic gas at 0°C and 1 atm. The system is subjected to the following processes, each of which is conducted reversibly:

(a) 10-fold increase in volume at constant temperature

(b) 100-fold adiabatic increase in pressure

(c) return to the initial state along a straight line path in the *P-V* diagram.

Calculate the work done by the system in step (c), and the total heat added to the system as a result of the cyclic process.

Chapter 3

Heat Capacity and Enthalpy— Auxiliary relations and applications

3.1 Heat Capacity of Ideal Gases

According to the *Kinetic Theory of Gases*, a monatomic gas molecule (He, Ar etc.) has three translational degrees of freedom. Each degree of freedom contributes $1/2R$ to C_V per mole according to the *Law of Equipartition of Energy*. Hence, C_V of one g-mole of a monatomic gas is equal to $3/2R$, which is approximately 12.5 J mol^{-1}K^{-1} Again, $C_P = C_V + R$, according to Eq. (2.42). Hence, C_P of a monatomic gas should be $5/2R$, and should be independent of temperature. This has been verified by experimental measurements.

A diatomic gas molecule (H$_2$, O$_2$, N$_2$ etc.) has two additional degrees of freedom arising from rotation. This gives a value of C_V as $5/2R$ and C_P as $7/2R$. This generally holds good at room temperature. However, experiments have demonstrated that C_P increases to almost $9/2R$ as temperature is raised. This is due to oscillation along the line joining the two atoms of the molecule. Such oscillation is associated with two additional vibrational degrees of freedom since it has both potential and kinetic energy.

For polyatomic gases, the situation is more complex. Hence, it is customary to employ experimentally measured (i.e. empirical) heat capacity vs. temperature relation for diatomic and polyatomic gas. Such relations have been proposed for solids and liquids as well, and are discussed in section 3.2.3.

3.2 Heat Capacity of Solids

In 1819, Dulong and Petit, on the basis of experimental data available then, proposed that molar heat capacities of all solid elements may be taken as $3R$ per°C. Further measurements subsequently showed this rule to be approximate and very limited in scope. Figure 3.1 presents C_V vs. T experimental data for a few solid elements at low temperature. It shows that:

(a) $C_V \rightarrow 0$ as $T \rightarrow 0$
(b) C_V increases with increase in T, but curves are different for different elements.
(c) In general, at high temperatures, $C_V \rightarrow 3R$ for some elements.

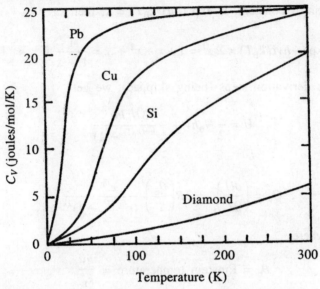

Fig. 3.1 Constant-volume molar heat capacities of Pb, Cu, Si and diamond as function of temperature.

3.2.1 Einstein's Theory of Heat Capacity of Solids

In the early 20th Century, Einstein applied quantum theory to explain experimental data on C_V of solids elements. Each atom in the crystal was assumed to behave as independent harmonic oscillator with oscillations along the three perpendicular directions. The energy of the ith level of harmonic oscillation is given as

$$\varepsilon_i = \left(i + \frac{1}{2}\right) h\nu \tag{3.1}$$

where ν is oscillation frequency and, h is Planck's constant. Again,

$$U = 3\Sigma n_i \varepsilon_i \tag{3.2}$$

where n_i = number of atoms in the ith energy level.

Combining Eqs. (3.1) and (3.2) and using the relation between n_i and Avogadro's number (N_0), we get

$$U = 3N_0 \; \Sigma \left(i + \frac{1}{2} \right) h\nu \left[\frac{\exp\left(-h\nu\left(i + \frac{1}{2}\right) \middle/ k_B T \right)}{\Sigma \exp\left(-h\nu\left(i + \frac{1}{2}\right) \middle/ k_B T \right)} \right] \qquad (3.3)$$

where k_B is Boltzmann's constant. If $\exp(-h\nu/k_B T) = x$, then

$$\underset{i}{\Sigma} \exp(-h\nu i/k_B T) = \Sigma \, x^i = 1 + x + x^2 + \dots = \frac{1}{1 - x} \text{ if } x \ll 1 \qquad (3.4)$$

Following Einstein's derivation steps (being skipped), we get

$$U = \frac{3}{2} N_0 h\nu + \frac{3N_0 h\nu}{(e^{h\nu/k_B T} - 1)} \qquad (3.5)$$

and

$$C_V = \left(\frac{\partial U}{\partial T} \right)_V = 3R \left(\frac{\theta_E}{T} \right)^2 \frac{e^{\theta_E/T}}{(e^{\theta_E/T} - 1)^2} \qquad (3.6)$$

(since $R = N_0 k_B$), where

$$\theta_E = \text{Einstein temperature} = \frac{h\nu}{k_B} \qquad (3.7)$$

3.2.2 Debye's Theory of Heat Capacity of Solids

Einstein's derivation was to be later modified by Debye who assumed a spectrum of vibration frequency in contrast to a single frequency of oscillation of Einstein. Debye arrived at a different equation in terms of *Debye temperature* (θ_D), where

$$\theta_D = \frac{h\nu_D}{k_B}, \; \nu_D = \nu_{max}$$

At very low temperatures (near $0°$ K), the Debye equation may be simplified as

$$C_V = 464.5(T/\theta_D)^3 \qquad (3.8)$$

The Debye equation gives better fit with experimental data, provided the best fitted values of θ_D are employed. Even then, it is not satisfactory. Moreover, it is concerned with elements only, that too in the crystalline state. Hence, due to its limited applicability, it is a normal practice to experimentally determine heat capacity vs. temperature relationships for solids and liquids.

3.2.3 Empirical C_P vs. T Relations

Experimental data consist of C_P as a function of temperature. *For solids and liquids, the PV term is very small. Hence, H is taken as equal to U, and C_P as equal to C_V.* In other words, no distinction is made between C_P and C_V so far as applications are concerned. It has been found that experimental C_P vs. T data for elements and compounds fit best with an equation of the type:

$$C_P = a + bT + cT^{-2} \tag{3.9}$$

where a, b, c are empirically fitted constants and differ from substance to substance.

The last term is the smallest and, therefore, often ignored. In some cases, such as liquid metals, both bT and cT^{-2} are usually ignored. Equation (3.9) is employed for correlation of experimental C_P data of diatomic and polyatomic gases as well. For alloys, where experimental data are not available, approximate estimates are done assuming heat capacity to vary linearly with atom fraction.

3.3 Enthalpy Changes

As already presented in Section 2.5, change of enthalpy during the course of a process (ΔH) is equal to the heat supplied to the system (q) at constant pressure, and that this is the basis for *process heat balance and thermal energy calculations.* Energy is costly. There is continuous attempt to save energy in all kinds of activity, including technological processing. By energy conservation, we essentially mean saving of fuels such as coal and petroleum. These are burnt to produce thermal energy, which is either utilized directly or converted into electrical energy.

The constant pressure restriction is mostly not important. H is not a function of P for ideal gases (see section 2.5.1). Energies of solids and liquids are hardly affected by some changes in pressure due to their very small molar volumes. In other words, the $V\,dP$ term in Eq. (2.25) is negligible. In most metallurgical and materials processing, gases are ideal and pressure is maximum a few atmospheres. Therefore, $\Delta H = q$ approximation is quite all right for almost all applications of our interest in the area of materials and metallurgical processing.

3.3.1 Classification of Enthalpy Changes

The absolute value of enthalpy of a substance is not known. All we can measure are changes of enthalpy. Again, enthalpy is an extensive property, and hence depends on the mass of the substance. *By Universal convention, standard thermochemical data books provide enthalpy changes per mole of a substance.*

Enthalpy changes occur due to various causes, and are classified as follows. Since heat is to be supplied to the substance or removed from it for enthalpy changes, the term *heat* is traditionally employed, although heat is not a state property, but energy in transition.

Sensible heat (i.e. sensible enthalpy)

Enthalpy change due to change of temperature of a substance is known as *sensible heat*. It is subdivided into:

(i) *Change in enthalpy without any change in state of aggregation of the substance.* It has already been presented in Eq. (2.36). *As a universal convention, thermochemical data books take sensible heat at 298 K (25°C) as zero for any substance.* Hence, sensible heat at temperature T per mole of a substance, is given as

$$H_T - H_{298} = \int_{298}^{T} C_P \, dT \qquad (3.10)$$

298 K (strictly 298.15) is known as *reference temperature*. Combining Eq. (3.10) with Eq. (3.9), we get

$$H_T - H_{298} = \int_{298}^{T} (a + bT + cT^{-2}) \, dT$$

$$= a(T - 298) + \frac{b}{2}(T^2 - 298^2) - c\left(\frac{1}{T} - \frac{1}{298}\right)$$

$$= AT + BT^2 - CT^{-1} + D \qquad (3.11)$$

where A, B, C, D are lumped parameters and functions of empirical constants a, b, c.

(ii) *Enthalpy changes due to changes in state of aggregation of substance.* These *are isothermal processes. By convention, enthalpy changes for all isothermal processes are designated by ΔH.* For example,

ΔH_m = enthalpy change of one mole of a solid due to melting (i.e. latent heat of fusion per mole)

ΔH_v = enthalpy change per mole of a liquid due to vaporization (i.e. latent heat of vaporization per mole)

Consider a *pure substance* A, which is undergoing the following changes during heating from 298 K to T K.

$$\begin{array}{ccccccc}
\text{A (solid)} & \to & \text{A (liquid)} & \to & \text{A (gas)} & \to & \text{A (gas)} \\
\text{at 298 K} & & \text{at } T_m & & \text{at } T_b & & \text{at } T
\end{array}$$

A is either a pure element or a pure compound. T_m and T_b are respectively, the melting and boiling points of A. Then,

$$H_T - H_{298} = \int_{298}^{T_m} C_P(\text{s}) \, dT + \Delta H_m + \int_{T_m}^{T_b} C_P(\text{l}) \, dT + \Delta H_v + \int_{T_b}^{T} C_P(\text{g}) \, dT \qquad (3.12)$$

where $C_P(\text{s})$, $C_P(\text{l})$ and $C_P(\text{g})$ are C_P's for solid, liquid and gaseous A, respectively. *If the substance is not a pure element or compound, then it does not have a single melting or boiling point, and hence Eq. (3.12) is not applicable.*

Sometimes an element or a compound undergoes *phase transformation* at the solid state. Then these terms are also to be added. An example is iron which exhibits the following structures at the solid state.

Phase	Structure	Temperature range (K) of stability
α	Body-centred cubic (BCC)	up to 1033
β	Nonmagnetic BCC	1033–1186
γ	Face-centred cubic (FCC)	1186–1665
δ	Body-centred cubic	1665–1809

δ-iron melts at 1809 K. Therefore, the sensible heat of iron at 2000 K is given as:

$$H_{2000} - H_{298} = \int_{298}^{1033} C_P(\alpha)\, dT + \Delta H_{tr}\,(\alpha \rightarrow \beta) + \int_{1033}^{1186} C_P(\beta)\, dT$$

$$+ \Delta H_{tr}(\beta \rightarrow \gamma) + \int_{1186}^{1665} C_P(\gamma)\, dT + \Delta H_{tr}\,(\gamma \rightarrow \delta) + \int_{1665}^{1809} C_P(\delta)\, dT$$

$$+ \Delta H_m + \int_{1809}^{2000} C_P(l)\, dT \tag{3.13}$$

EXAMPLE 3.1   Calculate ΔH and ΔU when one mole of water at 25°C at 1 atm pressure is converted into steam at 130°C and 2 atm.

Some relevant data (in J/mol) are now noted:
C_P of water = 75.3 per K, Latent heat of vaporization of water (ΔH_v) = 40.64 × 10³, C_P of steam between 100 – 130°C = 30.11 + 9.937 × 10⁻³T per K.

Solution $\Delta H = (\Delta H)_{25°C \rightarrow 130°C\ \text{at 1 atm}} + (\Delta H)_{1\ \text{atm} \rightarrow 2\ \text{atm at 130°C}}.$

Since steam is at low pressure, it may be treated as an ideal gas. Hence, (ΔH) steam for going from 1 atm to 2 atm at 130°C is zero, and

$$\Delta H = (\Delta H)_{25°C \rightarrow 130°C\ \text{(at 1 atm)}}$$

$$= \int_{25+273}^{100+273} C_P(\text{water})\, dT + \Delta H_v + \int_{100+273}^{130+273} C_P(\text{steam})\, dT$$

$$= 75.3 \times 75 + 40.64 \times 10^3 + \left[30.11\,T + \frac{9.937}{2} \times 10^{-3}T^2 \right]_{373}^{403}$$

$$= 47.31 \times 10^3 \text{ J/mol}$$

$$\Delta U = \Delta H - \Delta(PV) \qquad \text{[from Eq. (2.24)]}$$

Since PV is a state variable, $\Delta(PV)$ can be calculated along any path. Let us choose the following path:

$$\Delta(PV) = [\Delta(PV)]_{25°C \to 130°C \text{ at } 1 \text{ atm}} + [\Delta(PV)]_{1 \text{ atm} \to 2 \text{ atm at } 130°C}$$

$$\Delta(PV) \text{ at } 130°C = \Delta(RT) \text{ at } 130°C = 0 \text{ since } T \text{ is constant.}$$

Hence,

$$\Delta(PV) = [\Delta(PV)]_{25°C \to 130°C \text{ at } 1 \text{ atm}}$$

$$= [\Delta(PV)] \quad \text{for heating water from } 25°C \text{ to } 100°C$$

$$+ [\Delta(PV)] \quad \text{for vaporization of water at } 100°C$$

$$+ [\Delta(PV)] \quad \text{for heating steam from } 100°C \to 130°C \text{ at } 1 \text{ atm}$$

Since PV change is negligible for heating of water,

$$\Delta(PV) = P(V_s - V_w) + [\Delta(PV)]_{373}^{403}$$

where

$$P = 1 \text{ atm,}$$
$$V_s = \text{molar volume of steam at } 100°C \text{ (i.e. 373 K)}$$

$$= 22.4 \times 10^{-3} \times \frac{373}{273} = 30.6 \times 10^{-3} \text{ m}^3$$

$$V_w \approx 0 \text{ since it is very small.}$$

Hence,

$$P(V_s - V_w) = 1 \times 30.6 \times 10^{-3} \text{ m}^3 \text{ atm}$$

$$= 30.6 \times 10^{-3} \times 1.013 \times 10^5 = 3100 \text{ joules}$$

$$[\Delta(PV)]_{373}^{403} = [\Delta(RT)]_{373}^{403} = 8.314(403 - 373) = 250 \text{ joules}$$

Therefore,

$$\Delta U = 47.31 \times 10^3 - 3100 - 250 = 43.96 \times 10^3 \text{ J/mol}$$

Heat of reaction (ΔH)

This is the change of enthalpy that occurs when a reaction takes place. *By convention, the reaction is considered to be isothermal.* Consider the following reaction occurring at temperature, T:

$$\text{A (pure)} + \text{BC (pure)} = \text{AB (pure)} + \text{C (pure)} \qquad (3.14)$$

where A, B, C are elements and AB, BC are compounds. Then,

$$\Delta H(\text{at } T) = H_{AB}(\text{at } T) + H_C(\text{at } T) - H_A(\text{at } T) - H_{BC}(\text{at } T) \qquad (3.15)$$

where H_{AB}, H_C, H_A, H_{BC} are molar enthalpies of pure AB, C, A, BC, respectively.

Heat of mixing (ΔH_{mix})

The process of dissolution of a substance in a solvent (e.g. NaCl dissolution into water) is generalized now.

$$A \text{ (pure)} = A \text{ (in solution)} \tag{3.16}$$

This process is also accompanied by a change of enthalpy (ΔH_{mix}), where

$$\Delta H_{mix} = H \text{ (in solution)} - H_A \text{ (pure)} \tag{3.17}$$

Again, by convention, the process is assumed to be isothermal for thermodynamic calculations.

Some comments

For calculation of enthalpy changes, reactions, dissolutions and phase transformations have been assumed to occur isothermally. Since enthalpy is a state property, ΔH depends only on initial and final states, and not the path. The above assumption provides the most convenient path for calculation. Figure 2.1 has schematically shown variation of temperature with time when an isothermal process occurs. For a reversible isothermal process, the temperature remains constant all through. If the process is not reversible, then temperature at beginning and end of a process would be the same. In between, the temperature can vary significantly.

For calculation of enthalpy changes, *the isothermal processes are general (i.e. irreversible or reversible). Only the initial and final temperatures are assumed to be the same.* This is illustrated by Fig. 3.2 with the example of chemical reaction (3.14).

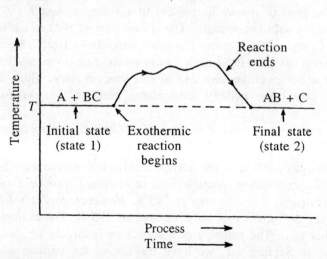

Fig. 3.2 Variation of temperature with progress of an irreversible exothermic isothermal reaction (schematic).

3.3.2 Conventions for ΔH and H

It has already been mentioned that thermochemical data sources provide $(H_T - H_{298})$ and ΔH values per mole (i.e. molar values).

The sign convention for ΔH

(a) Processes accompanied by liberation of heat are called *exothermic*. This happens if the enthalpy in the final state (state 2) is lower than that in the initial state (state 1), i.e. $H_2 < H_1$. So for the process

$$\text{State 1} \rightarrow \text{State 2},$$

we have

$$\Delta H = H_2 - H_1 < 0$$

Therefore, ΔH is negative. Figure 3.2 illustrates an exothermic reaction. *The opposite is an endothermic process which is characterized by absorption of heat and a positive value of ΔH.*

(b) Sensible heat (H) as well as ΔH for isothermal processes are functions of temperature. Hence, the temperature for which the value is being used is to be indicated. This is done in several ways. The first approach is to state the temperature separately, and not to use any symbol. Alternatively, these may be written as H_T, ΔH_T, $\Delta H(T)$ to denote that these are at temperature T K.

Standard state for enthalpy

A substance may be pure or it may be present in a solution. Again, the stable state of a pure substance changes with temperature. The stable state of H_2O is ice below 0°C, liquid water from 0–100°C, and a stable gas at 1 atm pressure above 100°C. Considering all these points, a standard state has been defined as *a pure element or compound at its stablest state at the temperature under consideration and at one atm pressure.* Thus, the standard state of H_2O at 50°C is pure water under 1 atmospheric pressure (i.e. *standard atmosphere* = 760 mm Hg).

By convention, enthalpies and enthalpy changes at standard state are denoted by superscript '0', e.g. H_T^0, ΔH_T^0.

As already mentioned, 298 K is the universal reference temperature for compilation of sensible heats. *By this convention, sensible heat at standard state of a substance (element or compound) is arbitrarily taken as zero at 298 K. However, this is solely for calculation of sensible heats, not ΔH^0 for a process occurring at 298 K.* For example, ΔH^0 (298) for reaction (3.14) is not zero. The enthalpy of a substance is largely or almost wholly due to its internal energy. In Section 2.1, we have mentioned the various sources of internal energy. Temperature, which is a measure of thermal potential, contributes to kinetic energy of atoms and molecules. This part comes under sensible heat. Other forms of internal energy are not very much influenced by temperature unless we go to extremely high temperature.

ΔH_f refers to heat effects when the reaction is concerned with formation of a compound from its elements. *Standard heat of formation* at a temperature T, i.e. $[\Delta H_f^0(T)]$ is nothing but ΔH_f at T, when elements and their compound are all at their respective standard states.

3.3.3 Experimental Measurements of Enthalpy Changes

Enthalpy changes of a substance are associated with heat effects—either absorption or evolution of heat. It has been stated in Chapter 1 that application of thermodynamics requires experimental data. Measurement of heat effects is one such requirement. It is done using a *calorimeter*. There are many designs of calorimeters.*

A calorimeter is an apparatus with known heat capacity. Therefore, from measurement of change of its temperature as a result of heat absorption/evolution due to the reaction/process, the quantity of heat can be estimated. For accurate measurements, as is the norm in thermochemistry, temperature changes have to be measured precisely (always within ±0.01°C and even lower). Mercury-in-glass thermometers, platinum resistance thermometers as well as thermocouples have been employed for this purpose.

The most widely used type of calorimeter is the *isoperibol Calorimeter*, in which the enclosure is held at constant temperature and the temperature of the substance undergoing a reaction/process is measured as a function of time. Since the calorimeter cannot be isolated from the surrounding completely, there will be some heat exchanges between them, causing errors. Such errors have to be eliminated through proper corrections.

In early days, calorimeters were essentially at room temperature. Even now that is true. However, over the last few decades, various designs of high temperature calorimeters going up to 2000°C have been developed and employed for measurements.

For measurement of specific heat of a substance, the latter is heated to a high temperature in a furnace and suddenly dropped into the calorimeter bath (typically water). From rise of bath temperature, specific heat is estimated.

3.3.4 Hess' Law

There are literally millions of chemical reactions. It is extremely difficult, if not impossible, to experimentally determine heats of so many reactions. Hess proposed a law, which enormously simplified the situation. Consider reaction (3.14). It may be considered as consisting of the following formation reactions at temperature T:

$$A(\text{pure}) + B(\text{pure}) = AB(\text{pure}); \quad \Delta H_{f,\,AB}^0(T) \tag{3.18}$$

$$B(\text{pure}) + C(\text{pure}) = BC(\text{pure}); \quad \Delta H_{f,\,BC}^0(T) \tag{3.19}$$

---

* See O. Kubaschewski, P.J. Spencer, and C.B. Alcock, *Materials Thermochemistry*, 6th ed., Pergammon Press, Oxford, 1993; see also earlier editions by O. Kubaschewski, E.Ll. Evans, and C.B. Alcock). Some standard data sources have been given in the Appendix to the book.

Subtraction of Eq. (3.19) from Eq. (3.18) yields Eq. (3.14), i.e.

$$A \text{ (pure)} + BC \text{ (pure)} = AB \text{ (pure)} + C \text{ (pure)}; \quad \Delta H^0(T) \tag{3.20}$$

According to Hess' Law, therefore,

$$\Delta H^0(T) = \Delta H^0_{f, AB}(T) - \Delta H^0_{f, BC}(T) \tag{3.21}$$

Hence, ΔH^0 of any reaction can be calculated, if the experimental values of ΔH^0_f of the relevant compounds are available.

Enthalpy is a state property. Hence, we may write

$$\int dH = 0, \text{ over a cycle (differential form)}$$

or

$$\sum_{cycle} \Delta H = 0 \quad \text{for finite stages} \tag{3.22}$$

For reaction (3.14), application of Eq. (3.22), which is the basis of Hess' Law, is illustrated by Fig. 3.3, where

$$\sum_{cycle} \Delta H^0 = \Delta H^0_{f, BC}(T) + \Delta H^0(T) - \Delta H^0_{f, AB}(T) = 0 \tag{3.23}$$

i.e. $\Delta H^0(T) = \Delta H^0_{f, AB}(T) - \Delta H^0_{f, BC}(T)$, which is Hess' Law [given in Eq. (3.21)].

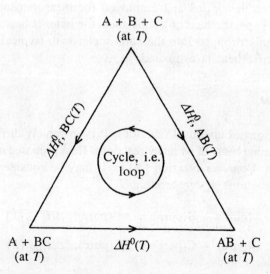

Fig. 3.3 Illustration of thermodynamic basis of Hess' Law; A, B, C, AB, BC are pure and one mole each.

Cyclic processes and the concept of cycle have been employed in thermodynamics and other subjects. In connection with computer programming, the term *loop* is more popular. Hence, the latter term is also being advocated now. Johnson and Stracher (1995)* have widely employed loops for problem solving.

EXAMPLE 3.2 Calculate ΔH^0_{298} for the reaction: $2CaO + SiO_2 = Ca_2SiO_4$

Given: ΔH^0_f for the formation of CaO, SiO_2 and Ca_2SiO_4 from their elements at 298 K are –634, –910.9 and –2305.3 kilojoules per mole (kJ mole^{-1}), respectively.

Solution

$$\Delta H^0_{298} = \Delta H^0_{f, 298}(Ca_2SiO_4) - 2\,\Delta H^0_{f, 298}(CaO) - \Delta H^0_{f, 298}(SiO_2)$$

$$= -2305.3 - [-2 \times 634 - 910.9] = -126.4 \text{ kJ per mole of reaction}$$

3.3.5 Kirchhoff's Law

ΔH for an isothermal process depends on the temperature at which the process occurs. The relationship of ΔH^0 of a reaction with temperature and other quantities is known as *Kirchhoff's Law*. Again, like Hess' law, it is based on the definition of enthalpy as a state property. The derivation is being illustrated for reaction (3.14) by application of the loop in Fig. 3.4.

$$\sum_{loop} \Delta H^0 = 0 = \Delta H^0_{T_1} + [(H^0_{T_2} - H^0_{T_1})_{AB} + (H^0_{T_2} - H^0_{T_1})_C]$$

$$-\Delta H^0_{T_2} - [(H^0_{T_2} - H^0_{T_1})_A + (H^0_{T_2} - H^0_{T_1})_{BC}] \tag{3.24}$$

or

$$\Delta H^0_{T_2} = \Delta H^0_{T_1} + \sum_{product}(H^0_{T_2} - H^0_{T_1}) - \sum_{reactant}(H^0_{T_2} - H^0_{T_1}) \tag{3.25}$$

Equation (3.25) is the generalized version of Kirchhoff's Law. If the reactants and products do not have any phase changes between T_1 and T_2, then Eq. (3.25) can be simplified as

$$\Delta H^0_{T_2} = \Delta H^0_{T_1} + \int_{T_1}^{T_2}\left[\sum_{product}C^0_P - \sum_{reactant}C^0_P\right]dT = \Delta H^0_{T_1} + \int_{T_1}^{T_2}\Delta C^0_P\,dT \tag{3.26}$$

If the reactants and products are not pure and not in their standard states, then also Kirchhoff's Law can be employed. However, the compositions of the solutions should not change with temperature. For melting and other phase transformations, the equation gets simplified. For example, for melting of solid,

$$\Delta H^0_m(T_2) = \Delta H^0_m(T_1) + \int_{T_1}^{T_2}[C_P(l) - C_P(s)]\,dT \tag{3.27}$$

---

*D.L. Johnson and G.B. Stracher, *Thermodynamic Loop Applications in Materials Systems,* Minerals, Metals Materials Society, Warrendale (Pennsylvania), 1995.

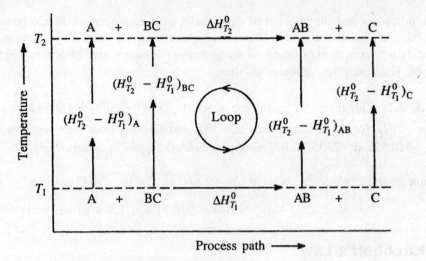

Fig. 3.4 Illustration for derivation of Kirchhoff's Law.

EXAMPLE 3.3 Calculate the adiabatic flame temperature when C_2H_4 gas is ignited at 298 K with stoichiometric amount of oxygen.

Given: (i) Heat of combustion (ΔH) of C_2H_4 at 298 K is -1411.3 kJ per mole of C_2H_4 (with gaseous reference state of H_2O at 298 K).

(ii) C_P of gases (J/mol/K)

CO_2 : $26.76 + 42.26 \times 10^{-3} \, T$

H_2O : $30.2 + 9.93 \times 10^{-3} \, T$

O_2 : $25.5 + 13.61 \times 10^{-3} \, T$

Solution If a fuel is assumed to burn under adiabatic condition (i.e. no heat loss to surrounding), then the resulting temperature of the products of combustion is known as *adiabatic flame temperature* (T_a). The consequent heat balance equation may be written as: Sensible heats of reactants at 298 K + heat released due to combustion at 298 K = heat required to raise temperature of products from 298 K to T_a. The sensible heats of reactants (i.e. C_2H_4 and O_2) at 298 K are zero by definition of sensible heat. The combustion reaction is:

$$C_2H_4 \,(g) + 3O_2 \,(g) = 2CO_2 \,(g) + 2H_2O(g) \qquad \text{(E.3.1)}$$

Therefore, the heat balance equation (per mole C_2H_4) is

$$1411.3 \times 10^3 + 0 = \int_{298}^{T_a} [2C_P(CO_2) + 2C_P\,(H_2O)] \, dT \qquad \text{(E.3.2)}$$

Substituting the values and rearranging the terms in Eq. (E.3.2), and on integration, we get

$$1411.3 \times 10^3 = 113.9 \, (T_a - 298) + \frac{104.4 \times 10^{-3}}{2} \, (T_a^2 - 298^2) \tag{E.3.3}$$

Equation (E.3.3) is a quadratic equation. By rearranging the terms, it becomes

$$52.19 \times 10^{-3} T_a^2 + 113.93 T_a - 1450 \times 10^3 = 0 \tag{E.3.4}$$

The valid solution is

$$T_a = \frac{-113.93 + [(113.93)^2 + 4 \times 52.19 \times 1450]^{1/2}}{2 \times 52.19 \times 10^{-3}}$$

$$= 4290 \text{ K}$$

[*Note:* H_2O is a stable liquid at 298 K. If that is employed as reference state at 298 K, the value of ΔH for combustion is to be adjusted as: $\Delta H = -1411.3 - 2\Delta H_v$, where ΔH_v is the molar heat of vaporization of H_2O. On the RHS also $2\Delta H_v$ is to be subtracted. Thus, these will automatically get cancelled, and the answer will be the same.]

EXAMPLE 3.4 An adiabatic vessel contains 1 kg of liquid aluminium at 700°C. Some Cr_2O_3 at 25°C was added into it, which caused the following exothermic reaction to occur:

$$2Al(l) + Cr_2O_3(s) = 2Cr(s) + Al_2O_3(s)$$

The final temperature was 1000°C, and the final product mixture consisted of $Al_2O_3(s)$, $Cr(s)$, and some unreacted $Cr_2O_3(s)$. Calculate the weight of Cr_2O_3 addition.

Solution Heat balance (with 298 K as reference temperature):

Sensible heat of reactants + heat released in reaction at 298 K
    = sensible heat of products (since adiabatic)

(a) Since Cr_2O_3 was added at 298 K, its sensible heat is zero. Hence,

Sensible heat of reactants = sensible heat of liquid Al at 973 K

$$= \frac{1000}{27} \left[\int_{298}^{932} C_P \, [\text{Al (s)}] \, dT + \Delta H_m \, (\text{Al}) + \int_{932}^{973} C_P \, [\text{Al (l)}] \, dT \right] \tag{E.3.5}$$

Since the atomic mass of Al = 27, and M.P. of Al = 932 K.
Noting

$$C_P[\text{Al (s)}] = 20.67 + 12.38 \times 10^{-3} \, T \text{ J/mol/K}, \qquad C_P[\text{Al (l)}] = 29.29 \text{ J/mol/K},$$

and

$$\Delta H_m(\text{Al}) = 10.46 \times 10^3 \text{ J/mol},$$

we have sensible heat of reactants = 10.96×10^5 joules

(b) Heat of reaction at 298

$$= \Delta H_{298}^0 = \Delta H_f^0 [Al_2O_3(s)]_{298} - \Delta H_f^0 [Cr_2O_3(s)]_{298}$$

$$= -16.74 \times 10^5 + 11.17 \times 10^5 = -5.57 \times 10^5 \text{ joules}$$

Hence,

Heat released $= 5.57 \times 10^5$ joules/mol

Therefore,

Total heat released $= \dfrac{n_{Al}}{2} \times 5.57 \times 10^5$

$$= 103.15 \times 10^5 \text{ joules}$$

where

n_{Al} = initial no. of g. moles of Al

$$= \dfrac{1000}{27} = 37.04$$

(c) The product has excess Cr_2O_3, but no excess Al. Hence,

$$n_{Cr_2O_3} \text{ (added)} = x > \frac{1}{2} n_{Al}$$

So,

$$n_{Al_2O_3} \text{ (in product)} = \frac{1}{2} n_{Al} = \frac{37.04}{2} = 18.52$$

$$n_{Cr} \text{ (in product)} = n_{Al} = 37.04$$

Therefore,

$$n_{Cr_2O_3} \text{ (in product)} = x - \frac{1}{2} n_{Al} = x - 18.52$$

The sensible heat of products at 1000°C (1273 K) is

$$\int_{298}^{1273} \{18.52\ C_P[Al_2O_3(s)] + 37.04\ C_P[Cr(s)] + (x - 18.52)]C_P[Cr_2O_3(s)]\} dT$$

(E.3.6)

$C_P = a + bT + cT^{-2}$ J/mol/K, with the following values of a, b, c:

	a	$b \times 10^3$	$c \times 10^{-5}$
Al_2O_3	106.6	17.78	−28.53
Cr	24.43	9.87	−3.68
Cr_2O_3	119.4	9.2	−15.65

Putting in the values in Eq. (E.3.6), and carrying out integration, we obtain

$$\text{Sensible heat of products} = 1.23 \times 10^5 \, x + 9.58 \times 10^5$$

Hence, the heat balance is

$$10.96 \times 10^5 + 103.15 \times 10^5 = 1.23 \times 10^5 x + 9.58 \times 10^5$$

i.e. $\qquad 10.96 + 103.15 - 9.58 = 1.23x \quad$ or $\quad x = 85$ g-moles

or weight of Cr_2O_3 added $= 85 \times (2 \times 52 + 3 \times 16) \times 10^{-3}$ kg $= 13.0$ kg.

3.3.6 Concluding Remarks

Some topics, concepts and conventions introduced in this chapter, such as standard state, isothermal process, loop application of state properties leading to Hess' Law and Kirchhoff's Law, would be used again for other important state properties, which will be introduced later.

Books on chemistry deal with chemical bonds. These bonds hold the atoms together in a material. The energy associated with a bond is known as *bond energy*. During the formation of a compound from elements, bonds are formed. This causes lowering of energy of the material and liberation of heat (i.e. ΔH is negative). Dissociation of a compound requires extra energy and, hence, heat is absorbed (i.e. ΔH is positive). Therefore, ΔH is often explained in terms of changes in bond energies. This is a simple way to look at enthalpy changes of a substance in chemical thermodynamics. However, it is only an approximation.

3.4 Summary

1. For ideal monatomic gases and ideal diatomic gases at not too high temperature, molar heat capacities at constant volume (C_V) and at constant pressure (C_P) are given by the Kinetic Theory of Gases.

2. For all other substances, the values of C_P as a function of temperature are determined experimentally, and are expressed by the relation

$$C_P = a + bT + cT^{-2}$$

   where a, b and c are empirical constants.

3. For solids and liquids, C_P and C_V are approximately the same.

4. Enthalpy changes are due to
   (a) change of temperature of a substance, and
   (b) other reactions and processes, e.g. phase transformations (including melting and vapourization), reaction, and mixing; these are all treated as isothermal processes.

5. Sensible heat (i.e. sensible enthalpy) of a substance is denoted as $H_T - H_{298}$, 298 K being the Universal reference temperature adopted by thermochemical data books. For isothermal processes, enthalpy change is denoted by ΔH.

6. At standard states of substances, H is designated as H^0, e.g. $H_T^0 - H_{298}^0$, ΔH^0

7. Calorimeters are employed for measurement of specific heats and enthalpy changes.

8. Hess' Law allows estimation of ΔH^0 of any reaction from ΔH_f^0 (i.e. heats of formation) of the compounds involved.

9. Kirchhoff's Law allows estimation of ΔH^0 of a reaction/process at any temperature from its value at another temperature.

10. Both Hess' and Kirchhoff's Laws are based on the fact that enthalpy is a state property.

PROBLEMS

3.1 Assuming no heat is lost to the surroundings, calculate the amount of heat required to just melt 1 kg of Pb initially at 15°C.

3.2 Consider oxidation of tungsten carbide (WC) as follows:

$$WC(s) + 5/2\ O_2(g) = WO_3\ (s) + CO_2(g), \quad \Delta H_{298}^0 = -1195.8 \text{ kJ mol}^{-1}$$

Calculate ΔH_f^0 of 1 mole of WC from elements at 298 K by Hess' Law.

3.3 Calculate the standard molar enthalpy of formation of titanium carbide (TiC) at 1200 K.

3.4 Calculate ΔH_{1800}^0 per mole of the reaction:

$$Mn\ (l) + \frac{1}{2}\ O_2\,(g) = MnO\ (s)$$

Note the following transformation steps for Mn from 298 K:

$$Mn(\alpha) \rightarrow Mn(\beta) \rightarrow Mn(\gamma) \rightarrow Mn(\delta) \rightarrow Mn(l),$$

3.5 Calculate ΔH^0 and ΔU^0 for the following reaction at 1000 K:

$$3FeO\ (s) + \frac{1}{2}\ O_2\ (g) = Fe_3O_4\,(s)$$

3.6 Which of the following processes releases more heat at 1000 K and by how much?
(a) Oxidation of graphite to CO
(b) Oxidation of diamond to CO

3.7 Calculate ΔH^0 per mole of the following reaction, occurring at 900 K.

$$2Al\,(s)\ +\ Cr_2O_3\,(s)\ =\ Al_2O_3\,(s)\ +\ 2Cr\,(s)$$

3.8 Calculate the adiabatic flame temperature for combustion of the following gas mixture: 25% CO, 12.5% CO_2, and 62.5% N_2 (by volume). The initial temperatures of the gas and air are 298 K. The theoretical amount of air was used. Assume that air contains 21% by volume O_2, rest N_2.

3.9 Calculate the heat of reaction at 1200 K per mole for the following reaction:

$$ZnO\,(s)\ +\ C\,(graphite)\ =\ Zn\,(g)\ +\ CO\,(g)$$

[*Note:* **From this chapter onwards, for problem solving, data are to be picked up from the Data Tables in Appendix, as required.**]

Chapter 4

Second Law of Thermodynamics and Entropy

4.1 Introduction

The First Law of Thermodynamics merely states that *if a process occurs*, an exact equivalence exists amongst various forms of energy changes. It provides no information regarding the feasibility of the process. The Second Law of Thermodynamics provides means to predict whether a particular process/reaction would take place under certain specified conditions, and thus is of great importance. Another important aspect of the Second Law, which is really fundamental to the problem enunciated above, is concerned with conversion of heat into work.

4.1.1 Various Statements of the Second Law

The Second Law of Thermodynamics has been stated in several equivalent forms since the time it was enunciated, implicitly by Carnot (1824), and explicitly first by Clausius (1850) and later independently by Kelvin (1851). According to Clausius and Kelvin, and as reformulated by Max Planck, the following statement is significant:

It is impossible to construct a cyclic engine that can convert heat from a reservoir at a uniform temperature completely into mechanical energy without leaving any effect elsewhere.

The engine must be cyclic, i.e. it returns to its initial position (i.e. state) after each stroke or revolution.

Some alternative statements are:

1. Heat absorbed at any one temperature cannot be completely transformed into work without leaving some change in the system or its surroundings.
2. *Spontaneous processes* are not thermodynamically reversible.

A spontaneous process occurs without external intervention of any kind. Examples are flow of heat from higher to lower temperature, diffusion of a species from higher concentration to lower concentration, mixing, and acid-base reactions. All natural processes, i.e. processes occurring in nature without external intervention, are spontaneous. In a refrigerator, heat is pumped out from lower to higher temperature by an artificial device. This device consumes electrical energy provided from outside.

4.2 Carnot Cycle

4.2.1 Introduction

A device utilizing heat to generate mechanical work and operating in a cycle is called a *heat engine*. The substance contained in it is called the *working substance*. Its operating feature is illustrated in Fig. 4.1.

$$\text{Efficiency of the engine } (\eta) = \frac{\text{work obtained per cycle}}{\text{heat absorbed per cycle}} = \frac{W}{q_2} \tag{4.1}$$

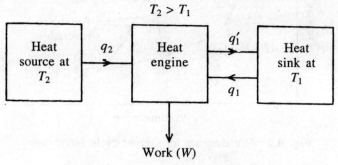

Fig. 4.1 Heat engine (schematic).

From the First Law of Thermodynamics,

$$\sum_{\text{cycle}} \Delta U = 0 = \text{net heat absorbed} - \text{work done}$$

$$= (q_2 - q_1') - W = (q_2 + q_1) - W \tag{4.2}$$

(*Note:* q denotes heat absorbed by the system, i.e. the engine; hence by convention $q_1' = -q_1$)

From Eqs. (4.1) and (4.2),

$$\eta = \frac{W}{q_2} = \frac{q_2 + q_1}{q_2} \tag{4.3}$$

S. Carnot (1824) conceptualized an ideal heat engine which operates reversibly between a heat source of constant temperature T_2 and heat sink of constant temperature T_1. Such an ideal reversible cycle is known as *Carnot Cycle* (*CS*), which laid the *foundation* of the Second Law of Thermodynamics.

It may be emphasized here that, for reversibility, not only the process should be carried out very slowly, but there should be no friction between the cylinder and the piston since frictional heat dissipation is irreversible.

4.2.2 Carnot Cycle with Ideal Gas as Working Substance

Figure 4.2 shows the Carnot cycle. It consists of four stages: two isothermal and two adiabatic. The temperature of isothermal stage $1 \rightarrow 2$ is T_2, and that of $3 \rightarrow 4$ is T_1. $T_2 > T_1$. All stages are reversible.

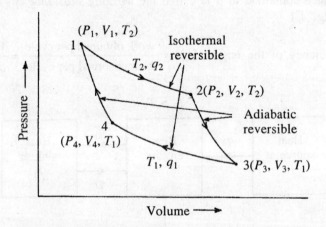

Fig. 4.2 *P-V diagram for Carnot cycle (schematic).*

Stage 1 → 2 (isothermal)

$\Delta U = 0$ for isothermal expansion/contraction of an ideal gas (section 2.5.1). Hence, from the First Law, and Eq. (2.13),

$$q_{1 \rightarrow 2} = q_2 = W_{1 \rightarrow 2} = nRT_2 \ln (V_2/V_1) \qquad (4.4)$$

where n = No. of moles of gas.

Stage 2 → 3 (adiabatic)

$$q_{2 \rightarrow 3} = 0; \text{ hence,}$$

$$W_{2 \rightarrow 3} = - (\Delta U)_{2 \rightarrow 3} = - nC_V(T_1 - T_2) \qquad (4.5)$$

Stage 3 → 4 (isothermal)

$$q_1 = W_{3 \to 4} = nRT_1 \ln (V_4/V_3) \tag{4.6}$$

Stage 4 → 1 (adiabatic)

$$W_{4 \to 1} = -(\Delta U)_{4 \to 1} = nC_v(T_1 - T_2) \tag{4.7}$$

Total work done in the cycle

$$W = W_{1 \to 2} + W_{2 \to 3} + W_{3 \to 4} + W_{4 \to 1}$$

$$= n \left[RT_2 \ln \left(\frac{V_2}{V_1} \right) - C_V(T_1 - T_2) + RT_1 \ln \left(\frac{V_4}{V_3} \right) + C_V(T_1 - T_2) \right] \tag{4.8}$$

Hence,

$$W = nR \left[T_2 \ln \left(\frac{V_2}{V_1} \right) + T_1 \ln \left(\frac{V_4}{V_3} \right) \right] \tag{4.9}$$

For adiabatic expansions, from Eq. (2.48),

$$\left(\frac{V_3}{V_2} \right)^{\gamma - 1} = \frac{T_2}{T_1} \tag{4.10}$$

$$\left(\frac{V_4}{V_1} \right)^{\gamma - 1} = \frac{T_2}{T_1} \tag{4.11}$$

Hence,

$$\frac{V_3}{V_2} = \frac{V_4}{V_1}, \text{ i.e. } \frac{V_4}{V_3} = \frac{V_1}{V_2} \tag{4.12}$$

Combining Eqs. (4.9) and (4.12), we get

$$W = nR \ln \left(\frac{V_2}{V_1} \right) (T_2 - T_1) \tag{4.13}$$

From Eqs. (4.13) and (4.4),

$$\eta = \frac{W}{q_2} = \frac{T_2 - T_1}{T_2} \tag{4.14}$$

Equation (4.14) shows that the efficiency of this reversible cycle depends only on T_2 and T_1.

4.2.3 Efficiency of Generalized Carnot Cycle

In a generalized Carnot cycle, the working substance is any thermodynamic substance. It is not restricted to ideal gas only. Section 1.3.3 has defined a thermodynamic substance which is at thermodynamic equilibrium and obeys Eq. (1.3). Since a reversible process proceeds through equilibrium states, it essentially has a working thermodynamic substance.

One of the statements of Generalized Carnot Theorem is:

All reversible cycles operating between the same upper and lower temperatures must have the same efficiency. The proof of this statement can be given in the following way. Figure 4.3 shows two engines working in cycles between the same heat reservoirs.

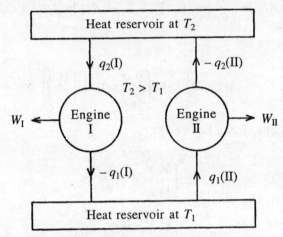

Fig. 4.3 A hypothetical system for proof of Carnot theorem.

Engine I is working with ideal gas and transferring heat from reservoir at T_2 to that at T_1. Engine II is working with another working substance and transferring heat in the reverse direction (i.e. $T_1 \rightarrow T_2$). Both engines are reversible.

Assume that

$$\text{Efficiency of engine I, } \eta_I > \text{efficiency of engine II, } \eta_{II} \qquad (4.15)$$

Then,

$$\eta_I = \frac{W_I}{q_2(I)} = \frac{q_2(I) + q_1(I)}{q_2(I)} \qquad (4.16)$$

$$\eta_{II} = \frac{W_{II}}{q_2(II)} = \frac{q_2(II) + q_1(II)}{q_2(II)} \qquad (4.17)$$

Assume further that

$$q_2(I) = -q_2(II) \qquad (4.18)$$

From Eqs. (4.16)–(4.18),

$$W_I + W_{II} = q_2(I) + q_1(I) + q_2(II) + q_1(II)$$

$$= q_1(I) + q_1(II) \tag{4.19}$$

Again,

$$\eta_I - \eta_{II} = \frac{W_I}{q_2(I)} - \frac{W_{II}}{(-q_2(I))} = \frac{W_I + W_{II}}{q_2(I)} > 0$$

or

$$W_I + W_{II} > 0 \tag{4.20}$$

Therefore, combining Eqs. (4.19) and (4.20), we obtain

$$W_I + W_{II} = q_1(I) + q_1(II) > 0 \tag{4.21}$$

In view of Eq. (4.18), there is no net heat flow into or out of reservoir at T_2. Since W_{II} is negative, $(W_I + W_{II})$ is the net work done by engines I and II. Similarly, $q_1(I)$ and $q_1(II)$ have opposite signs, $q_1(I)$ is negative and $q_1(II)$ is positive. Therefore, $[q_1(I) + q_1(II)]$ is the net heat transferred from the reservoir at T_1.

These mean that all the heat taken out of reservoir at T_1 is being converted into work, without leaving any change elsewhere (i.e. reservoir at T_2). This is not possible according to the Second Law of Thermodynamics. Hence, η_I cannot be different from η_{II}, and so,

$$\eta_I = \eta_{II} \tag{4.22}$$

Since engine I works with ideal gas, η_I is given by Eq. (4.14). Therefore,

$$\eta_{II} = \eta_I = \frac{T_2 - T_1}{T_2} \tag{4.23}$$

EXAMPLE 4.1 Two bodies (1 and 2) of equal heat capacity and initial temperatures T_1 and T_2 form an adiabatically closed system. What would be the final temperature if one lets the system come to thermal equilibrium (i) freely, and (ii) reversibly? What is the maximum work that can be obtained from the system?

Solution (i) Since the bodies are in adiabatic enclosure, no heat is exchanged with the surrounding. Therefore,

$$2mC_PT_f = mC_PT_1 + mC_PT_2 \tag{E.4.1}$$

where m is mass of a body and mC_P its heat capacity, and T_f is final temperature after attainment of thermal equilibrium.

From Eq. (E.4.1),

$$T_f = \frac{T_1 + T_2}{2}$$

(ii) Reversible attainment of thermal equilibrium is not possible without intervention of a reversible engine which absorbs heat from the body at higher temperature (say, at initially T_1) and rejects a part of it to the body at lower temperature (say, initially at T_2) reversibly.

Let T_a, T_b be instantaneous temperatures of body 1 and 2 at any instant. Let T_a change to $T_a + dT_a$, and T_b change to $T_b + dT_b$ due to transfer of an infinitesimal quantity of heat from body 1 to body 2. Now, at any instant,

$$\text{Efficiency of the reversible engine } (\eta) = \frac{T_a - T_b}{T_a} \qquad \text{(E.4.2)}$$

Again,

$$\frac{\text{Work done}}{\text{Heat absorbed}} = \frac{-mC_P(dT_a + dT_b)}{-mC_P\, dT_a} = \frac{dT_a + dT_b}{dT_a} \qquad \text{(E.4.3)}$$

(*Note*: dT_a is negative and dT_b is positive).

Combining Eqs. (E.4.2) and (E.4.3), we get

$$1 - \frac{T_b}{T_a} = 1 + \frac{dT_b}{dT_a}, \text{ i.e. } \frac{dT_b}{T_b} = -\frac{dT_a}{T_a} \qquad \text{(E.4.4)}$$

Integrating Eq. (E.4.4) between initial and final states, we get

$$\int_{T_2}^{T_f} \frac{dT_b}{T_b} = -\int_{T_1}^{T_f} \frac{dT_a}{T_a}, \text{ i.e. } \ln\frac{T_f}{T_2} = \ln\frac{T_1}{T_f} \qquad \text{(E.4.5)}$$

or

$$T_f = (T_1 T_2)^{1/2}$$

Work done by the reversible engine (W_{rev}) is the maximum work that can be obtained from the system.

Since the system is adiabatic,

$$\text{Total heat converted into work} = W_{\text{rev}} = mC_P(T_1 - T_f) - mC_P(T_f - T_2)$$

$$= mC_P[(T_1 + T_2) - 2(T_1 T_2)^{1/2}]$$

$$= mC_P(\sqrt{T_1} - \sqrt{T_2})^2$$

4.2.4 Thermodynamic Temperature Scale

In Carnot cycle for ideal gas, the temperature T was the absolute temperature as arrived at from Ideal Gas Laws. Since the efficiency of all reversible engines operating in cycle between the same temperatures are the same, Kelvin proposed that the absolute temperature scale be based on this thermodynamic conclusion. He called it the *thermodynamic temperature scale*. However, it was made the same as that based on ideal gas by fixing the freezing temperature of pure water (0°C) as 273.16 and that of boiling water as 373.16. In his honour, the name of the unit is known as *Kelvin* (K).

4.3 Entropy

4.3.1 Entropy—A state property

For a Carnot cycle (CS), Eqs. (4.3) and (4.14) yield

$$\frac{q_2 + q_1}{q_2} = \frac{T_2 - T_1}{T_2} \tag{4.24}$$

i.e.

$$\frac{q_1}{T_1} + \frac{q_2}{T_2} = 0 \tag{4.25}$$

for two isothermal stages. For adiabatic stages, $q = 0$. Hence for the entire cycle,

$$\sum_{\text{cycle}} \frac{q}{T} = 0 \tag{4.26}$$

Consider any arbitrary cycle. If it is reversible, then it may be regarded as consisting of a large number of tiny CS, as shown in Fig. 4.4. A → B →A is the arbitrary reversible cycle. The dashed lines divide it into many small Carnot cycles. The arrows indicate path directions for the small CS. In an interior cycle, its own path directions and those of its surrounding neighbours are opposite. Hence they cancel one another. Consider another CS at periphery. Here, the peripherial arrows are not opposed. Therefore, the net uncancelled lines of all the Carnot cycles are shown by solid lines.

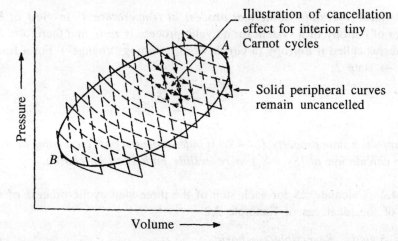

Fig. 4.4 *P-V* diagram for an arbitrary reversible cycle, divided into several tiny Carnot cycles.

For the net cycle described by the solid lines,

$$\Sigma \frac{q}{T} = 0 \tag{4.27}$$

since it is the sum of several Carnot cycles. If CS are made infinitesimally small (i.e. differential CS), then the solid lines would merge with the cycle ABA as the limiting case, and Eq. (4.27) may be written as

$$\int_{cycle} \frac{\delta q}{T} = 0 \tag{4.28}$$

for the cycle ABA. Since it is a reversible cycle, heat is absorbed reversibly. Hence Eq. (4.28) may be rewritten as

$$\int_{cycle} \frac{\delta q_{rev}}{T} = 0 \tag{4.29}$$

or

$$\int_{cycle} dS = 0 \tag{4.30}$$

where

$$dS = \frac{\delta q_{rev}}{T} \tag{4.31}$$

for an infinitesimal reversible isothermal process at temperature T. In view of Eq. (4.30), the net change of entropy of a system for a cyclic process is zero, and therefore, *S* is a *state property*. Clausius called it *Entropy* (a Greek word, meaning "change"). For a finite process from state 1 → state 2,

$$\Delta S = S_2 - S_1 = \sum_{1 \to 2} \frac{\delta q_{rev}}{T} \tag{4.32}$$

Since entropy is a state property, $(S_2 - S_1)$ is independent of path (reversible or irreversible). However, for calculation of $(S_2 - S_1)$, a reversible path is to be chosen.

EXAMPLE 4.2   Calculate ΔS for each step of the three step cyclic process of expansion/compression of the ideal gas of Example 2.2.

Solution   *Step I: Reversible isochoric*

$$\Delta S_I = \int_{T_1}^{T_2} \frac{\delta q_{rev}}{T} dT = nC_V \int_{T_1}^{T_2} \frac{dT}{T} = 10.617 \times 10^3 \text{ J K}^{-1}$$

Step II: *Reversible adiabatic:* $\Delta S_{II} = 0$, since $q = 0$

Step III: *Reversible isothermal*

$$\Delta S_{III} = \frac{q}{T} = -\frac{3185 \times 10^3}{300} = -10.617 \times 10^3 \ \text{J/K}^{-1}$$

Hence,

$$(\Delta S)_{\text{cycle}} = \Delta S_I + \Delta S_{II} + \Delta S_{III} = 10^3 \ (10.617 - 10.617) = 0$$

This is due to the fact that entropy is a state property.

4.3.2 Calculation of Entropy Changes

Entropy is a state property. Hence the basis and procedure for calculations of entropy changes are similar to those for enthalpy changes, as discussed in Section 3.3. Hess' Law and Kirchhoff's Law are applicable here too.

A pure substance at its stablest state also constitutes standard state for entropy at that temperature which *is designated as* S_T^0 (i.e. with superscript 0). The entropy values for pure substances and many non-standard substances (i.e. solutions) are available in thermodynamic data sources. Again, these were basically obtained from experimental measurements. However, the following significant differences are to be noted between entropy and enthalpy:

1. Entropy changes are to be calculated only through a reversible path. This restriction is not there for any other state property, including enthalpy.
2. The absolute value of entropy of a substance (S) can be determined. Thermodynamic data sources provide the values of S_T^0 for pure substances. This is a consequence of the Third Law of Thermodynamics, which will be presented in Chapter 12. This is in contrast with energy where only changes are available in data sources.

The entropy changes are to be calculated in the following ways, which are analogous to those for enthalpy.

Entropy changes for temperature and phase changes

(a) Heating and cooling can be done reversibly by carrying them out very slowly. Therefore, for 1 mole:

At constant volume,

$$dS = \frac{\delta q_{\text{rev}}}{T} = \frac{C_v \ dT}{T} = C_v d \ln T \tag{4.33}$$

At constant pressure,

$$dS = \frac{\delta q_{\text{rev}}}{T} = \frac{C_P \ dT}{T} = C_P d \ln T \tag{4.34}$$

(b) *Phase transformations can be made to occur slowly and reversibly*. For a pure substance, reversible phase changes (i.e. melting, boiling etc.) at a constant pressure occurs at a constant temperature. Equation (4.31) is for an infinitesimal process at const T. It can be easily converted into a finite difference form at constant *T*. Then *dS* becomes ΔS and δq_{rev} becomes q_{rev} (i.e. ΔH). Therefore, for melting of a pure substance,

$$\Delta S_m^0 = \frac{\Delta H_m^0}{T_m} \tag{4.35}$$

For boiling of a pure substance,

$$\Delta S_v^0 = \frac{\Delta H_v^0}{T_b} \tag{4.36}$$

In general, for phase transformation of a pure substance,

$$\Delta S_{tr}^0 = \frac{\Delta H_{tr}^0}{T_{tr}} \tag{4.37}$$

(c) Combining the above, for substance A and for the process

$$
\begin{array}{cccccc}
A \text{ (solid)} & \rightarrow & A \text{ (liquid)} & \rightarrow & A \text{ (gas)} & \rightarrow & A \text{ (gas)} \\
\text{at 298 K} & & \text{at } T_m & & \text{at } T_b & & \text{at } T
\end{array}
$$

we have

$$S_T^0 - S_{298}^0 = \int_{298}^{T_m} \frac{C_P^0(s)}{T}\, dT + \Delta S_m^0 + \int_{T_m}^{T_b} \frac{C_P^0(l)}{T}\, dT + \Delta S_v^0 + \int_{T_b}^{T} \frac{C_P^0(g)}{T}\, dT \tag{4.38}$$

(d) For a solution, ΔS_m etc. depend on the composition of the solution as well. Moreover, equilibrium (i.e. reversible) transformation temperature keeps changing as transformation progresses (see Section 10.4). Therefore, simple equations like Eqs. (4.35)–(4.38) are to be modified.

EXAMPLE 4.3 Calculate entropy of one mole of liquid Pb at 1000 K from that at 298 K.

Solution

$$S_{Pb}^0(1000) = S_{Pb}^0(298) + \int_{298}^{T_{m(Pb)}} \frac{C_P[Pb(s)]}{T}\, dT + \frac{\Delta H_m^0(Pb)}{T_m(Pb)} + \int_{T_m(Pb)}^{1000} \frac{C_P[Pb(l)]}{T}\, dT \tag{E.4.6}$$

on the basis of Eq. (4.38), where $T_m(Pb)$ = melting point of Pb = 600 K.
Substituting other values from data sources into Eq. (E.4.6), we obtain

$$S_{Pb}^0(1000) = 64.9 + \int_{298}^{600} \frac{[23.6 + 9.75 \times 10^{-3}\, T]}{T}\, dT + \frac{4810}{600} + \int_{600}^{1000} \frac{[32.4 - 3.1 \times 10^{-3}\, T]}{T}\, dT$$

$$= 64.9 + 23.6 \ln \frac{600}{298} + 9.75 \times 10^{-3}(600 - 298)$$

$$+ 8.017 + 32.4 \ln \frac{1000}{600} - 3.1 \times 10^{-3}(1000 - 600)$$

$$= 107.69 \text{ J K}^{-1} \text{ mol}^{-1}.$$

Entropy changes for reaction (ΔS) and mixing (ΔS_{mix})

By nature, reactions and mixing are irreversible processes. Like enthalpy changes, ΔS and ΔS_{mix} are to be considered only for isothermal processes. However, they cannot be calculated from ΔH and ΔH_{mix} using equations of the type (4.35)–(4.37) in view of irreversibility. For example, consider reaction (3.14) occurring isothermally at temperature T. For this,

$$\Delta S_T^0 = S_{AB}^0 \text{ (at } T) + S_C^0 \text{ (at } T) - S_A^0 \text{ (at } T) - S_{BC}^0 \text{ (at } T) \tag{4.39}$$

But taking $\Delta S_T^0 = \Delta H^0/T$ *will not be correct.*

The change of ΔS_T^0 with temperature can be calculated as

$$\Delta S_{T_2}^0 = \Delta S_{T_1}^0 + \sum_{\text{products}} (S_{T_2}^0 - S_{T_1}^0) - \sum_{\text{reactants}} (S_{T_2}^0 - S_{T_1}^0) \tag{4.40}$$

which is analogous to Eq. (3.25).

EXAMPLE 4.4 Calculate the entropy per mole of the following reaction at 1000 K from that at 298 K.

$$\text{Pb(l)} + \frac{1}{2} \text{O}_2(\text{g}) = \text{PbO(s)}$$

Solution From Eq. (4.40), ΔS^0 at 1000 K for the above reaction is related to ΔS_{298}^0 as follows:

$$\Delta S_{1000}^0 = \Delta S_{298}^0 + (S_{1000}^0 - S_{298}^0)_{\text{PbO}} - (S_{1000}^0 - S_{298}^0)_{\text{Pb}} - \frac{1}{2} (S_{1000}^0 - S_{298}^0)_{\text{O}_2} \tag{E.4.7}$$

$$\Delta S_{298}^0 = S_{298}^0(\text{PbO}) - S_{298}^0(\text{Pb}) - \frac{1}{2} S_{298}^0(\text{O}_2)$$

$$= 67.36 - 64.85 - \frac{1}{2} \times 205.0 \text{ (from data sources)}$$

$$= -100.0 \text{ J K}^{-1}$$

Noting that Pb melts at 600 K, and ΔH_m^0 of Pb = 4810 J/mol,

$$(S_{1000}^0 - S_{298}^0)_{PbO} - (S_{1000}^0 - S_{298}^0)_{Pb} - \frac{1}{2}(S_{1000}^0 - S_{298}^0)_{O_2}$$

$$= \int_{298}^{600} \frac{\Delta C_P'}{T}\, dT - \frac{4810}{600} + \int_{600}^{1000} \frac{\Delta C_P''}{T}\, dT \qquad \text{(E.4.8)}$$

where

$$\Delta C_P' = C_{P,\,PbO(s)} - C_{P,\,Pb(s)} - \frac{1}{2}C_{P,\,O_2(g)}$$

$$= -0.67 + 14.94 \times 10^{-3}T + 0.84 \times 10^5 T^{-2}$$

$$\Delta C_P'' = C_{P,\,PbO(s)} - C_{P,\,Pb(l)} - \frac{1}{2}C_{P,\,O_2(g)}$$

$$= -9.54 + 27.78 \times 10^{-3}T + 0.84 \times 10^5 T^{-2}$$

with C_P being molar heat capacity.

Integration of Eq. (E.4.8) and numerical calculations yield

$$\Delta S_{1000}^0 = -100.0 + 2.9 = -97.1 \text{ J K}^{-1}$$

Note: Analogous calculation can be performed on the basis of Kirchhoff's Law to find out heat of reaction (ΔH^0) at 1000 K for the above reaction. Since the procedure has been illustrated well in Chapter 3, we shall not deal with the details here. In fact, ΔH^0 is the molar heat of formation of PbO (ΔH_f^0), and ΔS^0 also is nothing but molar entropy of formation of PbO(ΔS_f^0).

$$\Delta H_{1000}^0 = \Delta H_{298}^0 + \int_{298}^{600} \Delta C_P'\, dT - \Delta H_m^0(Pb) + \int_{600}^{1000} \Delta C_P''\, dT \qquad \text{(E.4.9)}$$

Now, $\Delta H_{298}^0 = \Delta H_f^0(PbO)$ at 298 K $= -219.2 \times 10^3$ J/mol PbO (from data sources). Other data have already been listed above. Carrying out calculations, it is found that

$$\Delta H_{1000}^0 = -217 \times 10^3 \text{ J}$$

As stated in this section, chemical reactions are irreversible. Therefore, $\Delta S_T^0 \neq \Delta H_T^0/T$. This is confirmed, since here,

$$\frac{\Delta H_{1000}^0}{1000} = -\frac{217 \times 10^3}{1000} = -217 \text{ J/K},$$

and not equal to ΔS_{1000}^0.

4.4 Second Law and Entropy—Significance and consequences

4.4.1 General Comments

The statements of the Second Law of Thermodynamics are based on experience, and there is no direct proof for them. So many physicists in the 19th Century were reluctant to accept it as a part of physical science. The fundamental significance of entropy was also not known. General acceptance of the law by scientists required a few decades after Clausius' proposal of entropy in 1850. This came about when many experiments confirmed the correctness of predictions on the basis of the Second Law.

It took several more decades to fully establish the fundamental meaning of entropy qualitatively and quantitatively. This ended with the statistical interpretation of entropy and experimental verification of the same. The stepping stones were:

(a) Relation of thermal energy with random motions (translation, rotation and vibration) of atoms and molecules
(b) Statistical nature of macroscopic thermodynamic relations
(c) Concept of order and disorder
(d) Third Law of Thermodynamics

Developments in Physics in the 20th Century have shown that the basic laws and concepts, which constituted the foundation of physics upto 19th Century, are approximations only. These laws and concepts include:

• Law of Conservation of Matter
• Law of Conservation of Energy
• Separate entities for particle and wave

It has been established that matter can be converted into energy and vice versa. The elementary particles can be considered both as particle and wave. The only law proposed upto 19th Century, which is still held to be valid, is the Second Law of Thermodynamics.

4.4.2 Various Interpretations of Entropy

The various interpretations of entropy evolved gradually over a period of several decades. These are enumerated as follows:

1. $dS = \delta q_{rev}/T$ for an infinitesimal, isothermal, reversible process.
2. Entropy is Time's Arrow, i.e. a fundamental indicator of time.
3. Entropy has relation with heat not available for work.
4. Entropy is a measure of disorder of a system or a substance, which is under consideration.

The first interpretation has already been dealt with. The second and third interpretations will be arrived at following further discussions in this chapter. The fourth interpretation will be elaborated in relation to the Third Law and statistical interpretation of entropy in Chapter 12.

4.4.3 Entropy Changes for Reversible and Irreversible Processes

Conclusions regarding entropy changes for reversible and irreversible processes have been arrived at in various ways. Two procedures for derivation are presented here.

Derivation procedure I

Consider absorption of an infinitesimal quantity of heat (δq) by a closed system at a temperature T_{syst} from its surroundings. Assume that both system and surrounding have large heat capacities. Their temperatures remain essentially constant even after transfer of δq. Hence, the system absorbs heat isothermally and reversibly. The surrounding also gives out heat isothermally and reversibly. Thus,

$$(dS)_{system} = + \frac{\delta q}{T_{syst}} \tag{4.41}$$

$$(dS)_{surrounding} = - \frac{\delta q}{T_{surr}} \tag{4.42}$$

So,

$$(dS)_{syst + surr} = \delta q \left(\frac{1}{T_{syst}} - \frac{1}{T_{surr}} \right) \tag{4.43}$$

The reversible exchange of δq requires that the rate of heat transfer be very slow. This is possible if

$$T_{surr} \approx T_{syst} = T \tag{4.44}$$

Hence, for reversible heat exchange, from Eqs. (4.43) and (4.44),

$$(dS)_{syst + surr} = 0 \tag{4.45}$$

i.e.

$$(\Delta S)_{syst + surr} = 0 \quad \text{for finite process} \tag{4.46}$$

If T_{surr} is significantly higher than T_{syst}, then the heat transfer is irreversible. From

Eq. (4.43) since $\left(\dfrac{1}{T_{syst}} - \dfrac{1}{T_{surr}} \right) > 0$,

$$(dS)_{syst + surr} > 0 \tag{4.47}$$

i.e.

$$(\Delta S)_{syst + surr} > 0 \quad \text{for finite process} \tag{4.48}$$

Derivation procedure II

Consider the heat engine in Fig. 4.1 as the system and the two heat reservoirs as its surrounding.

That is, the engine and reservoirs are insulated from outside except for the heat and work interactions indicated in the figure. Since entropy is a state property, for one cycle,

$$(\Delta S)_{\text{engine}} = 0 \tag{4.49}$$

$$(\Delta S)_{\text{heat reservoirs}} = -\frac{q_2}{T_2} + \frac{q_1'}{T_1} = -\left(\frac{q_2}{T_2} + \frac{q_1}{T_1}\right) \tag{4.50}$$

For a reversible process, from Eq. (4.25),

$$(\Delta S)_{\text{heat reservoirs}} = 0 \tag{4.51}$$

Hence, from Eqs. (4.49) and (4.51), for a reversible process,

$$(\Delta S)_{\text{syst + surr}} = 0 \tag{4.52}$$

The cycle would be irreversible if a portion of the energy is dissipated as heat due to friction, rather than resulting into work. In that case, the efficiency of the engine would be lower than that of a reversible engine. In other words,

$$\eta_{\text{irr}} < \eta_{\text{rev}}, \text{ i.e. } \eta_{\text{irr}} < \frac{T_2 - T_1}{T_2} \tag{4.53}$$

i.e.

$$\frac{W}{q_2} = \frac{q_2 + q_1}{q_2} < \frac{T_2 - T_1}{T_2} \tag{4.54}$$

or

$$1 + \frac{q_1}{q_2} < 1 - \frac{T_1}{T_2} \tag{4.55}$$

i.e.

$$\frac{q_2}{T_2} + \frac{q_1}{T_1} < 0 \tag{4.56}$$

Combining Eq. (4.56) with Eq. (4.50), we get

$$(\Delta S)_{\text{heat resevoirs}} > 0 \tag{4.57}$$

For one cycle, $(\Delta S)_{\text{engine}} = 0$ [Eq. (4.49)]. Hence, for an irreversible cycle,

$$(\Delta S)_{\text{syst + surr}} > 0 \tag{4.58}$$

Here, the positive value of total entropy change for system and surrounding is essentially due to the irreversibility of the engine, which is making less heat available for work. Therefore, the larger the increase in $(\Delta S)_{\text{syst + surr}}$, the lesser the heat available for work. Thus, these two are interrelated. This provides explanation for the third interpretation of entropy, as described in Section 4.4.2.

EXAMPLE 4.5 Calculate the entropy change of the Universe in isothermal freezing of 1 g-mole of supercooled liquid gold at 1250 K, from the following data for gold.

$$T_m = 1336 \text{ K}, \quad \Delta H_m^0 \text{ (at 1336 K)} = 12.36 \times 10^3 \text{ J/mol}$$

$$C_P(\text{s}) = 23.68 + 5.19 \times 10^{-3}T \text{ J/mol/K}$$

$$C_P(\text{l}) = 29.29 \text{ J/mol/K}$$

Solution $\Delta S_{universe} = \Delta S_{syst} + \Delta S_{surr}$

(i) ΔS_{syst}

Actual freezing process is isothermal but irreversible, since it is not occurring at the equilibrium freezing temperature. But ΔS_{syst} is to be calculated only along a reversible path. It makes no difference as to which reversible path we choose since entropy is a state property. The simplest reversible path is

Liquid Au	$\xrightarrow{\text{heating}}$	liquid Au	$\xrightarrow{\text{freezing}}$	solid Au	$\xrightarrow{\text{cooling}}$	solid Au
At 1250 K		at 1336 K	at 1336 K	at 1336 K		at 1250 K
(state 1)						(state 2)

$$\Delta S_{syst} = S_2 - S_1 = \int_{1250}^{1336} \frac{C_P(\text{l})}{T} dT - \frac{\Delta H_m^0}{1336} + \int_{1336}^{1250} \frac{C_P(\text{s})}{T} dT$$

$$= \int_{1250}^{1336} \frac{29.69}{T} dT - \frac{12.36 \times 10^3}{1336} + \int_{1336}^{1250} \frac{(23.68 + 5.19 \times 10^{-3}\,T)}{T} dT$$

$$= -9.327 \text{ J K}^{-1}$$

(ii) ΔS_{surr}

Both the initial and final states of the system have been assumed to be 1250 K. This is possible only if the surrounding is at 1250 K, and it has infinite heat capacity (i.e. the surrounding absorbs heat released during freezing reversibly at 1250 K). Hence,

$$\Delta S_{surr} = -\frac{\Delta H_{syst}}{1250}$$

$$\Delta H_{syst} = H_2 - H_1 = \int_{1250}^{1336} 29.69 \, dT - 12.36 \times 10^3 + \int_{1336}^{1250} (23.68 + 5.19 \times 10^{-3}\,T) \, dT$$

$$= -12.46 \times 10^3 \text{ joules}$$

So,

$$\Delta S_{surr} = +\frac{12.46 \times 10^3}{1250} = 9.967 \text{ J/K}$$

or

$$\Delta S_{universe} = -9.327 + 9.967 = +0.64 \text{ J/K}$$

$\Delta S_{universe}$ is positive since the process is irreversible, as the derivation in Eq. (4.48) shows.

Note: In Example 4.4 also, the reaction is isothermal at 1000 K, but irreversible. Hence, the surrounding is at 1000 K and has large heat capacity. Hence the surrounding may be assumed to absorb heat released by the reaction reversibly at 1000 K.

$$\Delta S_{surr} = -\frac{\Delta H^0_{1000}}{1000} = +217 \text{ J K}^{-1}$$

i.e.

$$\Delta S_{univ} = \Delta S_{syst} + \Delta S_{surr} = 217 - 97.1 = +119.9 \text{ J K}^{-1}$$

ΔS_{univ} has a very high positive value for this reaction in contrast to that of freezing of supercooled gold. This is because a chemical reaction, in general, is more irreversible than irreversible freezing/melting.

4.4.4 Entropy Changes of Isolated Systems

An isolated system has no interaction with outside environment. From a thermodynamic point of view, therefore, the system contains the surroundings within itself. Therefore, from Eqs. (4.45) and (4.47), for an isolated system,

$$(dS)_{syst} = 0 \quad \text{for reversible process} \tag{4.59}$$

$$(dS)_{syst} > 0 \quad \text{for irreversible process} \tag{4.60}$$

The Universe is an isolated system. Moreover, spontaneous (i.e. irreversible) processes are occurring all the time. Hence, the entropy of the Universe is always increasing with the passage of time. There is no other property that behaves this way. Thus, entropy is an exception from others and is considered to be a fundamental indicator of time (Time's arrow). This explains the second interpretation of entropy (see section 4.4.2).

4.5 Summary

1. A heat engine converts heat into work and operates in cycle.
   Efficiency of an engine = $\eta = W/q_2$, where W is work done by the engine and q_2 is heat absorbed by it from the source, in one cycle.

2. For a reversible Carnot cycle operating between source temperature (T_2) and sink temperature (T_1), for any working substance,

$$\eta = \frac{T_2 - T_1}{T_2}$$

This is the basis for the thermodynamic temperature scale, proposed by Kelvin.

3. There exists a state property, entropy (S), defined as

$$dS = \frac{\delta q_{rev}}{T}$$

where δq_{rev} is an infinitesimal quantity of heat absorbed by the system reversibly at temperature T. For a finite process from state 1 to state 2,

$$S_2 - S_1 = \int_1^2 \frac{\delta q_{rev}}{T}$$

Hence, although S is a state property, change of entropy due to change of state has to be calculated along reversible path only. For an isothermal process at T, $\Delta S = q_{rev}/T$.

4. From the Third Law of Thermodynamics (not discussed yet), entropy of a perfectly ordered crystalline solid is zero at 0 K. This allows determination and tabulation of absolute values of entropy of substances at various temperatures.

5. Like enthalpy, entropy changes may be classified into:

   (a) changes due to change of temperature; and
   (b) changes due to isothermal processes, such as melting, phase transformations, reactions, and mixing.

   Also, Hess' and Kirchhoff's Laws are applicable.

6. $(\Delta S)_{syst+surr}$ $\begin{array}{l} = 0 \quad \text{for reversible process} \\ > 0 \quad \text{for irreversible process} \end{array}$

7. For an isolated system,

   $(\Delta S)_{system}$ $\begin{array}{l} = 0 \quad \text{for reversible processes} \\ > 0 \quad \text{for irreversible processes} \end{array}$

   since the system itself contains surrounding.

8. Efficiency of an irreversible engine (η_{irr}) is less than that of a reversible engine (η_{rev}).

PROBLEMS

4.1 Calculate ΔS for each stage for Problem 3.7.

4.2 The initial state of one mole of an ideal monatomic gas is $P = 10$ atm and $T = 300$ K. Calculate the entropy change in the gas for (a) reversible isothermal decrease in the pressure to 1 atm; (b) reversible adiabatic decrease in pressure to 1 atm; and (c) reversible constant volume decrease in pressure to 1 atm.

4.3 Calculate ΔS^0 at 298 K and 1 atm for the reaction:

$$W \text{ (s)} + O_2 \text{ (g)} = WO_2 \text{ (s)}$$

4.4 Calculate the standard entropy of formation of TiC from titanium and graphite at 1200 K.

4.5 Calculate ΔS for both system and surrounding per mole of the following isothermal process at 800 K: Al (l) = Al (s). Assume aluminium to be pure. Is the process irreversible?

4.6 Calculate ΔS^0 at 1000 K for the reaction in Problem 3.5. What is the value of ΔS of Universe?

4.7 One gram of liquid ThO_2 at 2900°C is mixed with 5 g of ThO_2 at 3400°C adiabatically.

(a) What is the final temperature?
(b) What is the entropy change of the system and the surrounding combined?
(c) Is the process spontaneous?

Assume C_P to be independent of temperature.

4.8 1 kg of a metallic powder initially at 25°C is mixed with 1 kg of the same liquid metal, initially at 150°C, in a container insulated from the surroundings. C_P of both liquid and powder are 4.0 J/g/K. Calculate the entropy change of the Universe for this process.

Chapter 5

Auxiliary Functions and Relations, Criteria for Equilibrium

5.1 Free Energy

As discussed in section 4.4.3, more the increase of entropy of the system during a process, less is the work done when a quantity of heat is supplied. The unit of entropy is joules per degree Kelvin ($J\ K^{-1}$). Hence the quantity TS has unit of energy (J). Approximately speaking, TS basically represents bound energy which cannot be utilized for doing work, but gets dissipated as heat. For knowing the extent of work the system is capable of doing, we ought to find out the *free energy, i.e. the energy available for work.*

The following two free energy functions are employed in thermodynamics, according to the names of scientists who proposed them:

Helmholtz Free Energy $A = U - TS$ (5.1)

Gibbs Free Energy $G = H - TS = U + PV - TS$ (5.2)

Since U, S, H are state properties and T is a state variable, A and G are also state properties by definition.

5.2 Combined Expressions of First and Second Laws of Thermodynamics

Assume: (i) closed system (i.e. fixed mass and fixed composition) and (ii) reversible process.

Then, from Eqs. (2.2) and (2.17),

$$dU = \delta q - \delta W = \delta q - P\ dV - \delta W' \tag{5.3}$$

From Eq. (4.31),

$$\delta q = T \, dS \tag{5.4}$$

Combining Eqs. (5.3) and (5.4), we get

$$dU = T \, dS - P \, dV - \delta W' \tag{5.5}$$

Again,

$$dH = dU + P \, dV + V \, dP \tag{2.25}$$

Combining Eqs. (5.5) and (2.25), we have

$$dH = (T \, dS - P \, dV - \delta W') + P \, dV + V \, dP$$

$$= T \, dS + V \, dP - \delta W' \tag{5.6}$$

Again, differentiating Eq. (5.1), we obtain

$$dA = dU - T \, dS - S \, dT \tag{5.7}$$

Combining Eq. (5.7) with Eq. (5.5), we get

$$dA = T \, dS - P\delta V - \delta W' - T \, dS - S \, dT$$

$$= - S \, dT - P \, dV - \delta W' \tag{5.8}$$

Differentiating Eq. (5.2), we obtain

$$dG = dH - T \, dS - S \, dT \tag{5.9}$$

Combining Eq. (5.9) with Eq. (5.6), we have

$$dG = (T \, dS + V \, dP - \delta W') - T \, dS - S \, dT$$

$$= - S \, dT + V \, dP - \delta W' \tag{5.10}$$

Equations (5.5), (5.6), (5.8) and (5.10) have been derived by combining the First and Second Laws of Thermodynamics with the assumption of reversible process, and for fixed mass and composition. If it is further assumed that the only work done is against pressure (i.e. $\delta W' = 0$), then these equations get further simplified into the following equations:

$$dU = T \, dS - P \, dV \tag{5.11}$$

$$dH = T \, dS + V \, dP \tag{5.12}$$

$$dA = - S \, dT - P \, dV \tag{5.13}$$

$$dG = - S \, dT + V \, dP \tag{5.14}$$

Equations (5.11)–(5.14) are four basic differential equations in thermodynamics, and are valid only under the assumptions stated above. *Closed and isolated systems have fixed mass and composition. The equations are applicable to open systems as well, provided we consider a fixed mass* (1 mole or 1 kg, for example), *and constant composition.*

5.3 Maxwell's Relations

Consider a variable Z which is a function of two variables *x, y*, i.e.

$$Z = f(x, y) \tag{5.15}$$

Then,

$$dZ = \left(\frac{\partial Z}{\partial x}\right)_y dx + \left(\frac{\partial Z}{\partial y}\right)_x dy = M\, dx + N\, dy \text{ (say)} \tag{5.16}$$

From the properties of differential equations,

$$\left(\frac{\delta M}{\delta y}\right)_x = \left(\frac{\partial N}{\partial x}\right)_y = \frac{\delta^2 Z}{\delta x\,\partial y} \tag{5.17}$$

Applying formula (5.17), we obtain from Eqs. (5.11)–(5.14) the following four differential equations:

$$\left(\frac{\delta T}{\delta V}\right)_S = -\left(\frac{\delta P}{\delta S}\right)_V \tag{5.18}$$

$$\left(\frac{\delta T}{\delta P}\right)_S = \left(\frac{\delta V}{\delta S}\right)_P \tag{5.19}$$

$$\left(\frac{\delta S}{\delta V}\right)_T = \left(\frac{\delta P}{\delta T}\right)_V \tag{5.20}$$

$$\left(\frac{\delta S}{\delta P}\right)_T = -\left(\frac{\delta V}{\delta T}\right)_P \tag{5.21}$$

Equations (5.18)–(5.21) are known as *Maxwell's relations*. These are the four basic equations. By combining these with other equations, it is possible to arrive at many other thermodynamic relations in the form of differential equations. They have been found to be very valuable in science and engineering.

Another property of function (5.15) is

$$\left(\frac{\delta Z}{\delta x}\right)_y \left(\frac{\delta x}{\delta y}\right)_z \left(\frac{\delta y}{\delta z}\right)_x = -1 \tag{5.22}$$

Some examples of derivation of useful equations employing some of the foregoing relations are being worked out now.

5.3.1 C_P, C_V **Relations for Thermodynamic Substances**

We have earlier derived

$$C_P - C_V = \left(\frac{\delta V}{\delta T}\right)_P \left[\left(\frac{\delta U}{\delta V}\right)_T + P\right] \tag{2.39}$$

From Eq. (5.11),

$$\left(\frac{\delta U}{\delta V}\right)_T = T\left(\frac{\delta S}{\delta V}\right)_T - P \tag{5.23}$$

Combining Eq. (5.23) with the Maxwell equation (5.20), we get

$$\left(\frac{\delta U}{\delta V}\right)_T = T\left(\frac{\delta P}{\delta T}\right)_V - P \tag{5.24}$$

From Eqs. (2.39) and (5.24),

$$C_P - C_V = T\left(\frac{\delta V}{\delta T}\right)_P \left(\frac{\delta P}{\delta T}\right)_V \tag{5.25}$$

Since $V = f(P, T)$, from Eq. (5.22), we have

$$\left(\frac{\delta V}{\delta P}\right)_T \left(\frac{\delta P}{\delta T}\right)_V \left(\frac{\delta T}{\delta V}\right)_P = -1 \tag{5.26}$$

From Eqs. (5.25) and (5.26),

$$C_P - C_V = -T\left(\frac{\delta V}{\delta T}\right)_P \left(\frac{\delta P}{\delta V}\right)_T \left(\frac{\delta V}{\delta T}\right)_P \tag{5.27}$$

Now, *isobaric coefficient of volumetric thermal expansion of a material* (α) is given as

$$\alpha = \frac{1}{V}\left(\frac{\delta V}{\delta T}\right)_P \tag{5.28}$$

Similarly, *isothermal compressibility of a material* (β) is given as

$$\beta = -\frac{1}{V}\left(\frac{\delta V}{\delta P}\right)_T \tag{5.29}$$

Combining Eqs. (5.27) to (5.29), we get

$$C_P - C_V = \frac{VT\alpha^2}{\beta} \tag{5.30}$$

Equation (5.30) is valid only for a thermodynamic substance since Eq. (5.11), which has been used for derivation, is applicable to *reversible processes* (i.e. *for a thermodynamic substance*).

5.3.2 Thermoelastic Effect

Consider an insulated (i.e. adiabatic) system. If a hydrostatic pressure P is applied, there will be a change in temperature of the substance. To make the process reversible, the pressure should be increased slowly. Also, deformation should be elastic. Under the above circumstances, we wish to find the relationship of $\left(\frac{\delta T}{\delta P}\right)_q$ with physical variables. The constant q restriction denotes that the system is adiabatic. When $\delta W' = 0$,

$$dU = \delta q - P\, dV$$

which is the same as Eq. (2.18). Again,

$$dU = \left(\frac{\delta U}{\delta V}\right)_T dV + \left(\frac{\delta U}{\delta T}\right)_V dT$$

which is the same as Eq. (2.7). Hence, at $\delta q = 0$,

$$dU = -P\, dV = \left(\frac{\delta U}{\delta V}\right)_T dV + \left(\frac{\delta U}{\delta T}\right)_V dT \tag{5.31}$$

Noting that $\left(\frac{\delta U}{\delta T}\right)_V = C_V$, and differentiating Eq. (5.31) with respect to P, Eq. (5.31) can be rewritten as

$$C_V\left(\frac{\delta T}{\delta P}\right)_q = -\left[\left(\frac{\delta U}{\delta V}\right)_T + P\right]\left(\frac{\delta V}{\delta P}\right)_q \tag{5.32}$$

Combining Eq. (5.32) with Eq. (2.39), we get

$$\left(\frac{\delta T}{\delta P}\right)_q = -\left[\frac{\dfrac{(C_P - C_V)}{C_V}}{\left(\dfrac{\delta V}{\delta T}\right)_P}\right]\left(\frac{\delta V}{\delta P}\right)_q \tag{5.33}$$

From Eqs. (5.33) and (5.30),

$$\left(\frac{\delta T}{\delta P}\right)_q = -\left[\frac{\dfrac{VT\alpha^2}{C_V\beta}}{\left(\dfrac{\delta V}{\delta T}\right)_P}\right]\left(\frac{\delta V}{\delta P}\right)_q \qquad (5.34)$$

Noting that $\alpha = \dfrac{1}{V}\left(\dfrac{\delta V}{\delta T}\right)_P$,

$$\left(\frac{\delta T}{\delta P}\right)_q = -\frac{T\alpha}{C_V\beta}\left(\frac{\delta V}{\delta P}\right)_q \qquad (5.35)$$

Again,

$$dV = \left(\frac{\delta V}{\delta T}\right)_P dT + \left(\frac{\delta V}{\delta P}\right)_T dP \qquad (5.36)$$

On the basis of definitions of α and β [see Eqs. (5.28) and (5.29)], we have

$$dV = V\alpha\, dT - V\beta\, dP \qquad (5.37)$$

or

$$\left(\frac{\delta V}{\delta P}\right)_q = V\alpha\left(\frac{\delta T}{\delta P}\right)_q - V\beta \qquad (5.38)$$

Substituting Eq. (5.38) into Eq. (5.35), we obtain

$$\left(\frac{\delta T}{\delta P}\right)_q = \frac{VT\alpha}{\left[C_V + \dfrac{VT\alpha^2}{\beta}\right]} = \frac{VT\alpha}{C_P} \qquad (5.39)$$

If we consider the effect of uniaxial tensile stress (σ), then P may be substituted by σ. Since the *coefficient of linear expansion is 1/3α, and the sign of tensile stress is opposite to pressure, which is compressive in nature*, Eq. (5.39) gets modified into

$$\left(\frac{\delta T}{\delta \sigma}\right)_q = -\frac{VT\alpha}{3C_P} \qquad (5.40)$$

for an isotropic substance.

Equation (5.40) gives the *thermoelastic coefficient* $\left(\dfrac{\delta T}{\delta \sigma}\right)_q$ in terms of physical variables. Experimental measurements have shown approximate agreement with this equation. The relationship between C_P and C_V, and the derivation of thermoelastic coefficient are two examples of the usefulness of Maxwell's relations. Many other relations have been obtained, and are listed in some standard texts.*

5.4 Criteria for Thermodynamic Equilibria

At equilibrium, there is no change in the system, i.e. it remains stable. Again, a reversible process occurs through a series of equilibrium stages. *Therefore, if we keep the independent variables on the right-hand side of Eqs. (5.11)–(5.14) fixed, then we arrive at the no change, i.e. equilibrium criteria.* Again, Eqs. (5.11)–(5.14) are for differential (i.e. infinitesimal) processes. The equilibrium criteria can be expressed in differential as well as in finite difference forms. Table 5.1 lists the criteria which constitute the equation given there viz. Eqs. [(5.41)–(5.44)].

Table 5.1 Equilibrium Criteria

Basis	Differential form	Finite difference form	Eq. Nos.
Eq. (5.11)	$(dU)_{S,V} = 0$	$(\Delta U)_{S,V} = 0$	(5.41)
Eq. (5.12)	$(dH)_{S,P} = 0$	$(\Delta H)_{S,P} = 0$	(5.42)
Eq. (5.13)	$(dA)_{T,V} = 0$	$(\Delta A)_{T,V} = 0$	(5.43)
Eq. (5.14)	$(dG)_{T,P} = 0$	$(\Delta G)_{T,P} = 0$	(5.44)

Since it is easy to maintain temperature and pressure constant, the Gibbs free energy criterion [Eq. (5.44)] is employed in chemical and metallurgical thermodynamics. However, in other areas, other criteria are also employed.

5.5 Criteria for Maximum Work

If $\delta W'$ is not zero, then other kinds of work are involved (e.g. electrical, magnetic, and surface etc.) besides $P\,dV$ type work. Then, from Eqs. (5.5), (5.6), (5.8) and (5.10), respectively, we arrive at the following equations:

$$(dU)_{S,V} = -\delta W', \text{ i.e. } (\Delta U)_{S,V} = -W' \tag{5.45}$$

$$(dH)_{S,P} = -\delta W', \text{ i.e. } (\Delta H)_{S,P} = -W' \tag{5.46}$$

$$(dA)_{T,V} = -\delta W', \text{ i.e. } (\Delta A)_{T,V} = -W' \tag{5.47}$$

$$(dG)_{T,P} = -\delta W', \text{ i.e. } (\Delta G)_{T,P} = -W' \tag{5.48}$$

---

*See, for example, R.A. Swalin, *Thermodynamics of Solids*, Wiley, New York, 1964, p. 32.

In Section 2.3, we have illustrated the concept of irreversible processes. Consider expansion of gas in the cylinder-piston system. For irreversible expansion, external pressure is significantly lower than the pressure in the system (say, $P - \Delta P$, where P is pressure inside). Then, $\delta W = (P - \Delta P)\, dV$. However, for a reversible process, it is $P\, dV$. *Therefore, the work done by a reversible process is maximum.*

Since we are considering reversible processes here, the work done would be maximum. Hence, W' in Eqs. (5.45)–(5.48) is also designated as $W'_{\max}$. Hence $(\Delta U)_{S,V}$ etc. denote the maximum capacity of the system to do work, and are loosely termed *as work content*. After the work is performed, the energy decreases. This is the significance of the negative sign. Application of these equations, especially Eq. (5.48), will be demonstrated in Chapters 13 and 14.

5.6 Criteria for Irreversible Processes

From Eqs. (4.3) and (4.24), for Carnot cycle,

$$\frac{W}{q_2} = \eta_{\text{rev}} = \frac{q_2 + q_1}{q_2} = \frac{T_2 - T_1}{T_2} \tag{5.49}$$

Suppose q_2 is absorbed irreversibly. Then, from Eqs. (4.53) and (5.49),

$$\eta_{\text{irr}} = \frac{q_2(\text{irr}) + q_1(\text{rev})}{q_2(\text{irr})} < \eta_{\text{rev}} \tag{5.50}$$

or

$$\frac{q_2(\text{irr}) + q_1(\text{rev})}{q_2(\text{irr})} < \frac{T_2 - T_1}{T_2} \tag{5.51}$$

Rearranging Eq. (5.51), we get

$$\frac{q_1(\text{rev})}{T_1} + \frac{q_2(\text{irr})}{T_2} < 0 \tag{5.52}$$

In general, for a cyclic process A $\rightarrow$ B $\rightarrow$ A, let A $\rightarrow$ B be irreversible and B $\rightarrow$ A be reversible. Then, with similar logic as given above,

$$\sum_{\text{A}\rightarrow\text{B}} \frac{\delta q(\text{irr})}{T} + \sum_{\text{B}\rightarrow\text{A}} \frac{\delta q(\text{rev})}{T} < 0 \tag{5.53}$$

Combining this equation with Eq. (4.32), we obtain

$$\sum_{\text{A}\rightarrow\text{B}} \frac{\delta q(\text{irr})}{T} + (S_{\text{A}} - S_{\text{B}}) < 0 \tag{5.54}$$

i.e.

$$S_B - S_A > \sum_{A \to B} \frac{\delta q (\text{irr})}{T} \qquad (5.55)$$

or

$$dS > \frac{\delta q (\text{irr})}{T} \qquad (5.56)$$

i.e. for an irreversible process,

$$\delta q < T \, dS \qquad (5.57)$$

Combining Eq. (5.57) with Eq. (2.2) and Eq. (2.17) and for $\delta W' = 0$, for irreversible process,

$$dU < T \, dS - P \, dV \qquad (5.58)$$

i.e.

$$(dU)_{S,V} < 0, \text{ i.e. } (\Delta U)_{S,V} < 0 \quad \text{for finite process} \qquad (5.59)$$

Similarly,

$$(dH)_{S,P} < 0 \quad \text{or} \quad (\Delta H)_{S,P} < 0 \qquad (5.60)$$

$$(dA)_{T,V} < 0 \quad \text{or} \quad (\Delta A)_{T,V} < 0 \qquad (5.61)$$

$$(dG)_{T,P} < 0 \quad \text{or} \quad (\Delta G)_{T,P} < 0 \qquad (5.62)$$

It should be noted that Eqs. (5.59)–(5.62) are qualitative in nature. Quantitative calculations are possible only for equilibrium [see Eqs. (5.41)–(5.44)] or for reversible processes, which proceed through a series of equilibrium stages.

5.7 Minimum Free Energy at Equilibrium

As stated in Section 5.4, it is relatively easy to maintain temperature and pressure constant. Hence the Gibbs free energy criteria are generally employed in chemical, metallurgical and materials thermodynamics. For this, J.W. Gibbs is considered as the father of chemical thermodynamics.

Let us recapitulate the criterion for equilibrium,

i.e.

$$(dG)_{T,P} = 0 \quad \text{or} \quad (\Delta G)_{T,P} = 0 \qquad (5.44)$$

and, for an irreversible process, viz.

$$(dG)_{T,P} < 0 \quad \text{or} \quad (\Delta G)_{T,P} < 0 \qquad (5.62)$$

For a process to occur spontaneously, Eq. (5.62) is to be satisfied. *Therefore, Eq. (5.62) provides the criterion for feasibility of a process.*

If, on the other hand, for a process,

$$(dG)_{T,P} > 0 \quad \text{or} \quad (\Delta G)_{T,P} > 0 \qquad (5.63)$$

then the process under consideration is not feasible. However, it may be noted that the process would tend to proceed spontaneously in the backward (i.e. reverse) direction, since for reverse process, the sign of $(dG)_{T,P}$ will be reverse, i.e. $(dG)_{T,P} < 0$.

The above considerations have led to the *minimum free energy criterion* for equilibrium. It is illustrated schematically in Fig. 5.1. It is evident from Fig. 5.1 that the system would tend towards state 2 either from the left or from the right. At state 2, $(dG)_{T,P} = 0$ since it is at minimum.

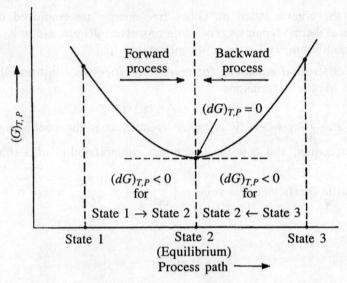

Fig. 5.1 Minimum free energy criterion for equilibrium.

Figure 5.1 constitutes the basis for the *free energy minimization technique* of searching equilibrium state of a system. This is the general method for complex equilibrium calculations. Figure 5.1 is two dimensional. For a three-dimensional arrangement, $(G)_{T,P}$ is a surface (like a bowl), and the minimum is at the bottom of the bowl.

5.8 Summary

1. Helmholtz free energy $A = U - TS$

   Gibbs free energy, $G = H - TS = U + PV - TS$

   By definition, free energy is a state property.

2. For a system of fixed mass and composition, and $\delta W' = 0$, the following criteria were arrived at by combining the First and Second Laws of Thermodynamics.

Reversible processes	Equilibrium	Spontaneous processes
$dU = T\, dS - P\, dV$	$(dU)_{S,V} = 0$	$(dU)_{S,V} < 0$
$dH = T\, dS + V\, dP$	$(dH)_{S,P} = 0$	$(dH)_{S,P} < 0$
$dA = -S\, dT - P\, dV$	$(dA)_{T,V} = 0$	$(dA)_{T,V} < 0$
$dG = -S\, dT + V\, dP$	$(dG)_{T,P} = 0$	$(dG)_{T,P} < 0$

Of these, the criteria based on Gibbs free energy are employed in chemical and metallurgical thermodynamics. For finite processes, dG etc. are to be replaced by ΔG etc. At equilibrium, $(G)_{T,P}$ is at its minimum.

3. The four differential equations for reversible processes constitute the basis for the four basic Maxwell's relations.

4. $C_P - C_V = \dfrac{VT\alpha^2}{\beta}$, where V is molar volume, α is the coefficient of volumetric thermal expansion, and β is the isothermal compressibility of a substance.

5. Thermoelastic coefficient of a solid $= \left(\dfrac{\delta T}{\delta \sigma}\right)_q = -\dfrac{VT\alpha}{3C_P}$, where σ is uniaxial tensile stress.

Chapter 6

Gibbs Free Energy and One-component Systems

6.1 Introductory Comments

6.1.1 Variables Influencing Gibbs Free Energy

As stated in Section 1.3, the state of a system decides its energy content. Again, the state depends on many variables. Free energy is no exception. This issue is again being elaborated for Gibbs free energy (G) with the intention of eventual understanding of its various applications in chemical thermodynamics.

Gibbs free energy G of a substance is a function of temperature and pressure. But the ΔG criterion is useful for application only at constant temperature (T) and pressure (P) for fixed mass of a system of constant composition (refer Chapter 5). Thus, let us first keep T and P constant, and consider the dependence of G of a substance on other variables.

Free energy is an extensive property and is proportional to mass (or mole) of the substance. Therefore, we should have a convention about the quantity of matter we are talking about. *By Universal convention in thermodynamics, the symbol G generally refers to Gibbs free energy per g-mole of a substance.* However, G also refers to free energy of the entire system. These will be specified in subsequent discussions. However, it is understood that if someone wishes to adopt some basis other than g-mole, it would not make any difference to conclusions in most cases.

The free energy of a substance *of fixed mass* also depends on

(a) chemical composition of the substance;
(b) state of aggregation of the substance (solid, liquid, gas or plasma);
(c) structure of the material, if it is a solid, such as crystal structure, microstructure, substructure, macrostructure;

(d) strengths of other energy fields (gravitational, electric, magnetic); and

(e) *specific surface area* (i.e. surface area per unit volume).

6.1.2 A Note on Subsequent Topics

It is also to be kept in mind that we are concerned with change of free energy (i.e. ΔG), and not its absolute value. In chemical thermodynamics, strengths of other energy fields [i.e. item (d) above] are assumed to remain constant during a process, and hence, its contribution to ΔG is negligible.

With the above discussion in mind, we shall consider the application of Gibbs free energy in thermodynamics of materials, which is based on chemical thermodynamics as stated in Chapter 1. It is to be further noted that quantitative application is possible only for thermodynamic equilibria in view of the criterion $(\Delta G)_{T,P} = 0$. Of course, this criterion is also applicable to reversible processes. However, it is beyond the scope of any introductory text on thermodynamics.

For irreversible processes, the criterion $(\Delta G)_{T,P} < 0$ is qualitative in nature. It can be employed for assessment of feasibility of a process only, and the same will be occasionally illustrated. Hence, for the reasons cited above, any text on chemical/metallurgical/material thermodynamics is basically concerned with the study of thermodynamic equilibria by which, we mean (a) phase equilibria in heterogeneous system, or (b) chemical equilibria in either homogeneous or heterogeneous systems.

In this chapter, we shall take up the simplest system of constant composition, as generally applicable to the one-component system (a pure element or pure compound). In subsequent chapters, more complex situations involving chemical and phase equilibria in multicomponent systems, where compositions of phases are variable, will be taken up one after another. Systems involving changes of electrical and surface energy, i.e. thermodynamics of electro-chemical cells and thermodynamics of surfaces, will be dealt with in Chapters 13 and 14.

As illustrated in Chapter 5, in solid state processes where volume remains approximately constant, internal energy (U) and Helmholtz free energy (A) criteria are more appropriate. In a way, these topics also belong to thermodynamics of materials but traditionally they are not covered in a subject like the present one.

6.2 Isothermal Processes

6.2.1 Some General Relations; Fugacity

For application of ΔG criterion, the process has to be isothermal. It can be reversible or irreversible. Only the initial and final temperatures should be the same (see Section 3.3).

Consider the process

$$\text{State 1 (at temp } T) \rightarrow \text{State 2 (at temp } T) \qquad (6.1)$$
$$(G_1, H_1, S_1) \qquad\qquad (G_2, H_2, S_2)$$

G_1, H_1, S_1 and G_2, H_2, S_2 are free energy, enthalpy and entropy values at states 1 and 2, respectively. Then,

$$\Delta G = G_2 - G_1, \qquad \Delta H = H_2 - H_1, \qquad \Delta S = S_2 - S_1 \tag{6.2}$$

Again, from the definition of G (Eq. 5.2),

$$\Delta G = G_2 - G_1 = (H_2 - TS_2) - (H_1 - TS_1)$$
$$= (H_2 - H_1) - T(S_2 - S_1)$$
$$= \Delta H - T\,\Delta S \tag{6.3}$$

In Eq. (6.1), the process from state 1 to state 2 may be any isothermal process viz. phase transformations, mixing, chemical reactions (see Section 3.3). Suppose the process is purely a change of pressure for a pure substance A, i.e.

$$A(P_1,\ T) \rightarrow A(P_2,\ T) \tag{6.4}$$

then what should be ΔG? If A is solid or liquid, $\Delta G \approx 0$, unless the change of pressure is very large since pressure does not have much effect on their energy (see Section 3.3). However, if A is a gas, then change of pressure would have a significant effect on G. From Eq. (5.14), at const T, $dG = V\,dP$. If A is an ideal gas, then

$$dG = V\,dP = (RT/P)\,dP = RT\,d(\ln P) \tag{6.5}$$

Integrating Eq. (6.5) from P_1 to P_2, for the process (6.4),

$$\Delta G = RT\,\ln\,(P_2/P_1) \tag{6.6}$$

The unit of pressure in chemical thermodynamics is one standard atmosphere (760 mm Hg), *simply designated as atm.*

If the gas is not ideal, then, in analogy with Eq. (6.5), a function known as fugacity *(f) has been defined as follows:*

$$dG = RT\,d(\ln f) \tag{6.7}$$

It may be noted here that the above definition of fugacity is a general one applicable to all thermodynamic substances, and not restricted to nonideal gases only.

EXAMPLE 6.1 The equilibrium vapour pressure of ice at $-10°C$ is 1.95 mmHg, and that of supercooled water at the same temperature is 2.149 mmHg. Calculate the free energy change accompanying the change of 1 mole of supercooled water to ice at $-10°C$.

Solution Free energy is a state property. Hence we can choose any path for calculation with the same result. Let us choose the following path:

$$\text{1 mole supercooled} \xrightarrow{\Delta G_1} \text{Water vapour} \xrightarrow{\Delta G_2} \text{Water vapour}$$
$$\text{water } (-10°C) \qquad (2.149 \text{ mm Hg}, -10°C) \ (1.95 \text{ mm Hg}, -10°C)$$
$$\downarrow \Delta G_3$$
$$\text{ice } (-10°C)$$

$$\Delta G = \Delta G_1 + \Delta G_2 + \Delta G_3$$

$$= 0 + RT \ln \frac{1.95}{2.149} + 0$$

$$= -212.5 \text{ J mol}^{-1} \quad \text{at } T = 263 \text{ K}$$

Note: ΔG_1 and ΔG_3 are zero because these steps are at equilibrium.

6.2.2 Standard State—Fugacity relationship

The standard state of a substance has already been defined in Section 3.3 in connection with enthalpy. The same statement is applicable to free energy, i.e. *the standard state is pure element or compound at 1 atm pressure and at its stablest state at the temperature under consideration (i.e. at ambient temperature). It may be stated further that it should be ideal, if it is gaseous.*

It has also been mentioned in Section 3.3 that the thermodynamic quantities at standard state of a substance are distinguished by superscript zero (H^0, S^0, G^0, ...).

Now, in Eq. (6.6), if $P_1 = 1$ atm and $P_2 = P$ (i.e. a variable pressure), and $G_2 = G$, then

$$\Delta G = G - G^0 = RT \ln P \tag{6.8}$$

Similarly, for nonideal gas, Eq. (6.7) may be integrated to yield

$$G - G^0 = RT \ln(f/f^0) \tag{6.9}$$

where f^0 is fugacity at standard state.

It is to be noted further that any gas tends to behave ideally if $P \to 0$. i.e.

$$\text{if } P \to 0, \text{ then } f \to P, \quad f/P \to 1 \tag{6.10}$$

6.2.3 Fugacity—Pressure Relationship in Gases

Gases tend to deviate from ideal behaviour more if pressure is increased and/or temperature is decreased. The permanent gases exhibit significant departure from ideality only at extremely low (sub-zero) temperatures and/or very high pressures. In metals and materials processing, these are rare. Most processing are at high temperature and moderate pressure. Hence departures from ideal behaviour are small.

For small deviations from ideality,

$$V = \frac{RT}{P} - \alpha \tag{6.11}$$

where α is a temperature dependent parameter. Hence, from Eqs. (6.7) and (6.11),

$$dG = V \, dP = RTd(\ln f) = RTd \ln P - \alpha \, dP \tag{6.12}$$

i.e.

$$d\left[\ln\left(\frac{f}{P}\right)\right] = -\frac{\alpha}{RT}\,dP \qquad (6.13)$$

Integrating Eq. (6.13), we get

$$\ln\left(\frac{f}{P}\right) = -\frac{\alpha P}{RT} + C \qquad (6.14)$$

where C is the integration constant. If integration is done from $P = 0$ to $P = P$, then at $P = 0$, $f/P = 1$, and hence, $C = 0$. This yields

$$f/P = \exp\left(-\frac{\alpha P}{RT}\right) \qquad (6.15)$$

Again, for small deviation from ideality, $\alpha P/RT \ll 1$. So,

$$\frac{f}{P} \approx 1 - \frac{\alpha P}{RT} \approx \frac{PV}{RT} \quad \text{[from Eq. (6.11)]} \qquad (6.16)$$

Again,

$$\frac{RT}{V} = P_i \qquad (6.17)$$

where P_i is pressure of ideal gas corresponding to the values of V and T in Eq. (6.16). Combining Eqs. (6.16) and (6.17), we get

$$\frac{f}{P} \approx \frac{P}{P_i} \qquad (6.18)$$

From experimental values of α, some values of f were calculated for H_2, O_2, CH_4 at $P = 50$ atm at a few temperatures. These are given in Table 6.1. It may be noted that the gases are fairly close to ideal, especially at 200°C. So at lower pressure, they may be assumed to be ideal without any significant error.

Table 6.1 Values of f at $P = 50$ atm

Gas	0°C	100°C	200°C
H_2	51.5	51.2	51.0
O_2	48.5	50.0	50.5
CH_4	45.2	48.5	50.0

6.3 Variation of ΔG with Temperature for an Isothermal Process

For this purpose, mass, composition and pressure are assumed to be constant. Only the temperature change is being considered.

6.3.1 Gibbs–Helmholtz (G–H) Equations

From Eq. (5.14),

$$\left(\frac{\delta G}{\delta T}\right)_P = -S \tag{6.19}$$

Combining Eq. (6.19) with Eq. (5.2), we get

$$G = H - TS = H + T\left(\frac{\delta G}{\delta T}\right)_P \tag{6.20}$$

Dividing LHS and RHS of Eq. (6.20) by T^2, we obtain

$$\frac{G}{T^2} = \frac{H}{T^2} + \frac{1}{T}\left(\frac{\delta G}{\delta T}\right)_P$$

i.e.

$$\frac{1}{T}\left(\frac{\delta G}{\delta T}\right)_P - \frac{G}{T^2} = -\frac{H}{T^2} \tag{6.21}$$

or

$$\left[\frac{\delta(G/T)}{\delta T}\right]_P = -\frac{H}{T^2} \tag{6.22}$$

Equation (6.22) is one form of G–H equation. The alternative form is derived through differentiation by parts as follows:

$$\left[\frac{\delta(G/T)}{\delta\left(\frac{1}{T}\right)}\right]_P = \frac{1}{T}\left[\frac{\delta G}{\delta(1/T)}\right]_P + G = \frac{1}{T}\left(\frac{\delta G}{\delta T}\right)_P \frac{dT}{d(1/T)} + G \tag{6.23}$$

So,

$$\left[\frac{\delta(G/T)}{\delta(1/T)}\right]_P = -T^2\frac{1}{T}\left(\frac{\delta G}{\delta T}\right)_P + G = -T^2\left[\frac{1}{T}\left(\frac{\delta G}{\delta T}\right)_P - \frac{G}{T^2}\right] \tag{6.24}$$

Combining Eq. (6.24) with Eq. (6.21), we get

$$\left[\frac{\delta(G/T)}{\delta(1/T)}\right]_P = -T^2\left(-\frac{H}{T^2}\right) = H \tag{6.25}$$

Equation (6.25) is the alternative form of G–H equation.

For the process, as in Eq. (6.1), *if dT, dP are common to both states 1 and 2, then*

$$dG = dG_2 - dG_1 = d(\Delta G)$$

$$= (V_2 - V_1)dP - (S_2 - S_1)\,dT$$

$$= \Delta V\,dP - \Delta S\,dT \tag{6.26}$$

So,

$$\left[\frac{\delta(\Delta G)}{\delta T}\right]_P = -\Delta S \tag{6.27}$$

Since Eq. (6.27) is analogous to Eq. (6.19), the following relations can be derived:

$$\left[\frac{\delta(\Delta G/T)}{\delta T}\right]_P = -\frac{\Delta H}{T^2} \tag{6.28}$$

$$\left[\frac{\delta(\Delta G/T)}{\delta(1/T)}\right]_P = \Delta H \tag{6.29}$$

where ΔG is change of Gibbs free energy for an isothermal process. Similarly, ΔV, ΔS, ΔH also refer to changes in V, S, H due to occurrence of the isothermal process.

6.4   Phase Equilibria in One-component (i.e. Unary) Systems

In Section 1.3.4, we have briefly discussed about classification and characterization of systems for application of thermodynamics. We have mentioned there that *a system is homogeneous if it consists of a single phase and heterogeneous if it has more than one phase.* For example, water and ice are different phases of the same compound, H_2O.

The question is: what is a phase? It may be stated that a phase is characterized by *phase boundary which is a surface where properties change abruptly (i.e. discontinuously).* As already discussed in Section 1.3.4 and Section 6.1, heterogeneous equilibria may be classified into:

1. Phase equilibria for nonreactive systems, where no chemical reaction occurs
2. Phase equilibria for reactive systems, which are accompanied by chemical reactions (By convention, it is mostly treated as chemical equilibria rather than phase equilibria).

Thus, by phase equilibria, we generally mean heterogeneous equilibria in a nonreactive system. In this section, we shall discuss about one-component system which consists of a single element or compound (also known as Unary system). The question of chemical reaction, of course, does not arise at all here, as undoubtedly it is a nonreactive system.

6.4.1 Clapeyron Equation

Consider the phase transformation of an element or compound (designated as A).

$$A \text{ (phase I)} = A \text{ (phase II)} \tag{6.30}$$

At equilibrium phase transformation temperature (T_{tr}) and pressure (P_{tr}),

$$\Delta G = G_A(II) - G_A(I) = 0 \tag{6.31}$$

where $G_A(I)$ and $G_A(II)$ are Gibbs free energy per mole of A in phases I and II, respectively. Since transformation occurs at constant T and P, $\Delta G = 0$, as per the criterion of equilibrium.

Now, let the pressure be changed to $P_{tr} + dP$. Then allow sufficient time for the phase equilibrium to be re-established at $P_{tr} + dP$. Then T_{tr} will also change to $T_{tr} + dT$, $G_A(I)$ will change to $G_A(I) + dG_A(I)$, $G_A(II)$ will change to $G_A(II) + dG_A(II)$, at the new equilibrium. As already stated in Chapter 5, these changes will be related to other variables, as

$$dG = V \, dP - S \, dT \tag{5.14}$$

Accordingly,

$$dG_A(I) = V_A(I) \, dP - S_A(I) \, dT \tag{6.32}$$

$$dG_A(II) = V_A(II) \, dP - S_A(II) \, dT \tag{6.33}$$

Again, at the new equilibrium,

$$\Delta G = [G_A(II) + dG_A(II)] - [G_A(I) + dG_A(I)] = 0 \tag{6.34}$$

From Eqs. (6.31) and (6.34),

$$dG_A(I) = dG_A(II) \tag{6.35}$$

Combining Eqs. (6.32), (6.33) and (6.35), we get

$$V_A(I) \, dP - S_A(I) \, dT = V_A(II) \, dP - S_A(II) \, dT \tag{6.36}$$

i.e.

$$\left(\frac{dT}{dP}\right)_{eq} = \frac{V_A(II) - V_A(I)}{S_A(II) - S_A(I)} = \frac{\Delta V_{tr}}{\Delta S_{tr}} \tag{6.37}$$

Again, from Chapter 4,

$$\Delta S_{tr} = \frac{\Delta H_{tr}}{T_{tr}} \tag{4.37}$$

Therefore,

$$\left(\frac{dT}{dP}\right)_{eq} = \frac{\Delta V_{tr} \cdot T_{tr}}{\Delta H_{tr}} \tag{6.38}$$

Equation (6.38) is known as the Clapeyron equation, and is applicable to phase transformations involving only condensed phases (i.e. solids and liquids). $\left(\dfrac{dT}{dP}\right)_{eq}$ is the rate of change of equilibrium phase transformation temperature with change of pressure. ΔV_{tr}, ΔS_{tr}, ΔH_{tr} are difference, respectively, of molar volume, entropy and enthalpy of A between phases II and I.

Change of pressure has negligible effect on ΔV_{tr} and ΔH_{tr}. Also, change of temperature is typically very small. Hence, *these are taken as constants* and equal to their values at the standard pressure of 1 atm., i.e. at standard states of A in both phases. As discussed in Section 3.3.2, *values at standard state are designated by superscript '0'.* Therefore, Eq. (6.38) may also be written as

$$\left(\frac{dT}{dP}\right)_{eq} = \frac{\Delta V_{tr}^0 \cdot T_{tr}^0}{\Delta H_{tr}^0} \tag{6.39}$$

For example, the Clapeyron equation for melting/freezing of a pure element or compound may be written as

$$\left(\frac{dT_m}{dP}\right)_{eq} = \frac{\Delta V_m^0 \cdot T_m^0}{\Delta H_m^0} \tag{6.40}$$

From Eq. (6.19), at constant pressure, $dG/dT = -S$. *According to the Third Law of Thermodynamics, entropy of a substance is always positive.* This makes the slope of dG/dT negative, and decreases free energy of a substance with increase of temperature. This is illustrated schematically in Fig. 6.1 for $G_A(I)$ and $G_A(II)$. It also shows that

$$\begin{aligned}
\Delta G &> 0 \quad \text{if } T < T_{tr} \\
&= 0 \quad \text{at } T = T_{tr} \\
&< 0 \quad \text{at } T > T_{tr}
\end{aligned} \tag{6.41}$$

Thus, phase II is stabler than phase I at $T > T_{tr}$, and reverse is true at $T < T_{tr}$.

The sign of $\left(\dfrac{dT}{dP}\right)_{eq}$ can be either positive or negative. Let us consider melting of solid. *During melting, a substance absorbs heat, i.e.* ΔH_m^0 *is positive.* For most solids, the volume per mole increases upon melting, i.e. ΔV_m^0 is positive. These make the sign of $\left(\dfrac{dT_m}{dP}\right)_{eq}$ positive as per Eq. (6.40), i.e. T_m increases with increase of pressure. *Exceptions to this are*

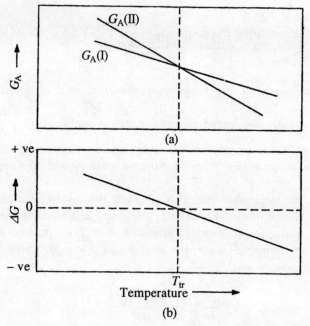

Fig. 6.1 Variation of: (a) G_A (I), G_A (II), and (b) ΔG, with temperature (schematic).

H_2O, *bismuth etc., where* ΔV_m^0 *is negative* and the contraction of volume occurs upon melting. T_m will decrease with increase of P. Figure 6.2 presents the phase diagram for H_2O; S, L

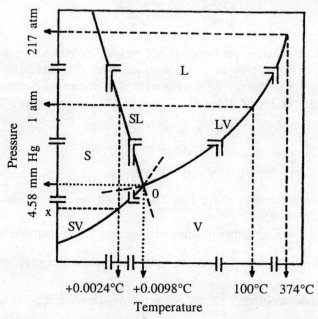

Fig. 6.2 Phase diagram for H_2O; dashed lines show metastable extensions. (P. Saha, in *Elements of Ceramic Science,* D. Ganguly and S. Kumar (Eds.), Indian Institute of Ceramics, Calcutta, 1982, p. 94.)

and V denote regions of stability of solid, liquid and vapour, respectively. Line SL is solid-liquid transition boundary obeying the Clapeyron equation. It may be noted that increasing P lowers T for SL.

EXAMPLE 6.2 Calculate the change of equilibrium transformation temperature of α-iron into γ-iron for raising pressure from 1 to 100 atm, with the help of the following data:

 (i) 910°C is the equilibrium transformation temperature for this at 1 atm pressure.
 (ii) At 910°C, $\Delta H^0_{\alpha \to \gamma} = 912$ J mol^{-1}, densities of α- and γ-iron are 7.571 and 7.633 g cm^{-3}, respectively.
 (iii) Ignore change of volume with pressure.

Solution

$$\left(\frac{dT}{dP} \right)_{eq} = \frac{\Delta V^0_{tr} \cdot T^0_{tr}}{\Delta H^0_{tr}} \tag{6.39}$$

$T^0_{tr} = 910 + 273 = 1183$ K.

Since the atomic mass of Fe is 55.85, the molar volume of $\alpha - Fe = \dfrac{55.85}{7.571} = 7.377$ cm^3 mol^{-1}, and that of $\gamma - Fe = \dfrac{55.85}{7.633} = 7.317$ cm^3 mol^{-1}. Hence,

$$\Delta V^0_{tr} = 7.317 - 7.377 = -0.06 \text{ cm}^3 \text{ mol}^{-1} = -0.06 \times 10^{-6} \text{ m}^3 \text{ mol}^{-1}$$

Again,

$$\Delta H^0_{tr} = 912 \text{ J mol}^{-1} = \frac{912 \times 82.06 \times 10^{-6}}{8.314} = 9.0 \times 10^{-3} \text{ m}^3 . \text{ atm/mol}$$

since $R = 82.06$ cm^3.atm/(mol. K). Substituting the values in Eq. (6.39), we get

$$\left(\frac{dT}{dP} \right)_{eq} = -\frac{0.06 \times 1183 \times 10^{-6}}{9.0 \times 10^{-3}} = -7.88 \times 10^{-3} \text{ K/atm}$$

Thus, change of T_{tr} (ΔT_{tr}) due to change of pressure from 1 atm to 100 atm is

$$\Delta T_{tr} = \frac{dT}{dP} (100 - 1) = -0.78 \text{ K}$$

EXAMPLE 6.3 Carbon has two allotropes, graphite and diamond. At 25°C and 1 atm pressure, graphite is the stable form. Calculate the minimum pressure which must be applied to graphite at 25°C in order to bring about its transformation to diamond.
 Given:
 (i) H^0_{298} (graphite) $- H^0_{298}$ (diamond) $= -1900$ J/mol

(ii) S_{298}^0 (J/mol.K) are 5.735 and 2.428 for graphite and diamond, respectively.

(iii) Densities of graphite and diamond at 25°C are 2.22 and 3.515 gm/cm^3, respectively.

Solution Transformation of graphite into diamond would be thermodynamically feasible, if for the process: graphite → diamond, $\Delta G < 0$. Now,

$$dG = V\, dP - S\, dT$$

$$dG = V\, dP, \text{ i.e. } d(\Delta G) = \Delta V \cdot dP \qquad \text{at constant } T, \qquad \text{(E.6.1)}$$

where $\Delta V = V_{\text{diamond}} - V_{\text{graphite}}$, where V is molar volume. This is how change of pressure affects relative stabilities of graphite and diamond.

At 1 atm and 25°C (i.e. 298 K),

$$\Delta G = G_{\text{dia}}^0 - G_{\text{gr}}^0 = (H_{\text{dia}}^0 - H_{\text{gr}}^0) - 298(S_{\text{dia}}^0 - S_{\text{gr}}^0)$$

$$= 1900 - 298\,(2.428 - 5.735)$$

$$= 2886 \text{ J mol}^{-1} = 2886 \times \frac{82.06 \times 10^{-6}}{8.314} = 0.0285 \text{ m}^3.\text{atm/mol}$$

At the minimum pressure required for transformation (P_{min}), $\Delta G = 0$. Hence, the integral form of Eq. (E.6.1) is

$$\int_{0.0285}^{0} d(\Delta G) = \int_{1}^{P_{\text{min}}} (\Delta V)\, dP \qquad \text{(E.6.2)}$$

Now,

$$\Delta V = 12 \times 10^{-6}\left(\frac{1}{3.515} - \frac{1}{2.22}\right) = -1.99 \times 10^{-6} \text{ m}^3.\text{mol}^{-1}$$

Solving Eq. (E.6.2), $P_{\text{min}} = 14.3 \times 10^3$ atm.

6.4.2 Phase Equilibria with Vapour: Clausius–Clapeyron Equation

The transformation of a liquid into its vapour is known as *vaporization*. The corresponding phase equilibrium is vaporization equilibrium. The transformation of a solid directly into its vapour is called *sublimation*, and its equilibrium is *sublimation equilibrium*. Both vaporization and sublimation equilibria will be discussed jointly since their basic equations are similar in form.

Since the molar volume of vapour (V_v) of a substance is orders of magnitude larger than those in condensed states (V_c), we have

$$\Delta V_{\text{tr}} = V_v - V_c \approx V_v \qquad \text{(6.42)}$$

Again, vaporization and sublimation can occur over a wide range of temperature. For example, water vapour co-exists at equilibrium with water from 0°C to 100°C. *The equilibrium vapour pressure (i.e. saturated vapour pressure) is a function of temperature.* As Figure 6.2 shows, it is about 5 mm Hg at 0°C and increases with rise in temperature. Therefore, ΔH_{tr} and T_{tr} are variables, and hence would be denoted as ΔH and T, respectively. In normal phase equilibria involving vapours, the total pressure in the system is not very large and is of the order of a few atmospheres at the most. Therefore, ΔH may be taken as equal to ΔH^0, i.e. ΔH at standard state of 1 atm.

With the above modifications, from Eqs. (6.39) and (6.42),

$$\left(\frac{dP}{dT}\right)_{eq} = \frac{\Delta H^0}{TV_v} \tag{6.43}$$

where ΔH^0 is either ΔH_v^0, i.e. *heat of vaporization* or ΔH_S^0, i.e. *heat of sublimation,* of a pure substance.

Assuming the vapour to behave as an ideal gas, we have

$$V_v = \frac{RT}{p_A^0} \tag{6.44}$$

where p_A^0 is vapour pressure of pure A in atm. Unit of gas constant R would depend on the unit of volume. Suppose V_v is in m^3, then the unit of R would be m^3.atm mol^{-1} K^{-1}. The assumption of vapour as ideal gas is reasonable for metallurgy and materials science since vapour pressures are typically very low.

From Eqs. (6.43) and (6.44),

$$\frac{1}{p_A^0}\left(\frac{dP}{dT}\right)_{eq} = \frac{\Delta H^0}{RT^2} \tag{6.45}$$

Imagine a closed vessel at temperature T, containing a solid or a liquid. If sufficient time is allowed, the vapour pressure inside the vessel will keep increasing till equilibrium is attained and it becomes constant. If the vessel is initially evacuated, then the total pressure P would be the same as p_A^0. On the other hand, if the vessel was not evacuated to start with, then $P > p_A^0$.

In either case, however, P should be substituted by p_A^0 since it is only p_A^0, which is the common pressure to both the vapour and the condensed phase. Hence, Eq. (6.45) should be further modified as

$$\frac{1}{p_A^0}\left(\frac{dp_A^0}{dT}\right)_{eq} = \frac{\Delta H^0}{RT^2} \tag{6.46}$$

i.e.

$$\frac{d(\ln p_A^0)_{eq}}{dT} = \frac{\Delta H^0}{RT^2} \tag{6.47}$$

An alternative form of Eq. (6.47) is

$$\frac{d(\ln p_A^0)_{eq}}{d(1/T)} = -\frac{\Delta H^0}{R} \tag{6.48}$$

Either Eq. (6.47) or Eq. (6.48) is Clausius-Clapeyron (CC) equation in differential form. The subscript "eq" on LHS can be omitted as long as we understand that we are referring to equilibrium vapour pressure in the CC equation.

6.4.3 Effect of Total Pressure on Vapour Pressure

The total pressure P in the system should have effect on the value of p_A^0 in the following ways:

(a) Properties of condensed phase depend on P.
(b) If the system is at a very high pressure (say, above 50 atm or so), then the vapour would not behave as an ideal gas (see section 6.2.3). In that case, p_A^0 should be replaced by fugacity of pure A (i.e. f_A^0). This effect can be neglected for change of P in normal ranges.

Although properties of condensed phase depend on P, they have negligible effect on vapour pressure, as the following analysis shows.

Consider the situation where liquid A is kept in a closed chamber at temperature T and inert gas pressure P. After some time, the pressure of vapour of A in the empty space would attain the equilibrium value (p_A^0). Now, let us change inert gas pressure P to $P + dP$ in the chamber. Then after some time, a new vapour-liquid equilibrium will be established with saturated vapour pressure as $p_A^0 + d\,p_A^0$.

For change of P to $P + dP$, at new equilibrium, dG of condensed phase, viz.

$$dG_{cond} = dG_{vap} \tag{6.49}$$

where dG refers to change of G as a result of change of P. At constant T,

$$dG_{cond} = V_{cond}\, dP, \qquad dG_{vap} = V_{vap}\, dp_A^0 \tag{6.50}$$

Combining Eqs. (6.49) and (6.50), we get

$$\frac{dp_A^0}{dP} = \frac{V_{cond}}{V_{vap}} \tag{6.51}$$

Since V_{vap} is approximately 10^3 to 10^5 times that of V_{cond} per mole of A, dp_A^0/dP is negligibly small, i.e. change of total pressure has negligible effect on p_A^0.

6.4.4 Integration of Clausius–Clapeyron Equation

ΔH^0 is a function of temperature. However, for many situations, ΔH^0 may be approximately taken as constant. Then integration of Eq. (6.48) leads to

$$\ln p_A^0 = -\frac{\Delta H^0}{RT} + I \tag{6.52}$$

where I is integration constant. At normal boiling point, $p_A^0 = 1$ atm, i.e. $\ln p_A^0 = 0$ and $T = T_b^0$. Substitution of these in Eq. (6.52) yields

$$\ln p_A^0 = -\frac{\Delta H^0}{R}\left(\frac{1}{T_b^0} - \frac{1}{T}\right) \tag{6.53}$$

Suppose we are considering vaporization of a liquid. From Eq. (3.26),

$$\Delta H_T^0 = \Delta H_{298}^0 + \int_{298}^T (\Delta C_P^0)\, dT \tag{6.54}$$

where

$$\Delta C_P^0 = C_P^0(v) - C_P^0(l) \tag{6.55}$$

In general, for both liquid A and vapour of A, Eq. (3.11) is applicable individually, i.e.

$$H_T^0 - H_{298}^0 = AT + BT^2 - CT^{-1} + D \tag{3.11}$$

where A, B, C, D are lumped parameters, and are empirical constants. Equation (3.11) has been derived from

$$H_T^0 - H_{298}^0 = \int_{298}^T C_P^0 \cdot dT = \int_{298}^T (a + bT + cT^{-2})\, dT \tag{6.56}$$

$$\Delta C_P^0 = (a + bT + cT^{-2})_{\text{vap}} - (a + bT + cT^{-2})_{\text{liq}}$$

$$= \Delta a + \Delta b \cdot T + \Delta c \cdot T^{-2} \tag{6.57}$$

On this basis, we may rewrite Eq. (3.11) as

$$\Delta H^0 = \Delta H_T^0 = \Delta H_{298}^0 + \Delta A \cdot T + \Delta B \cdot T^2 - \Delta C \cdot T^{-1} + \Delta D' \tag{6.58}$$

$$= \Delta A \cdot T + \Delta B \cdot T^2 - \Delta C \cdot T^{-1} + \Delta D$$

since $\Delta H_{298}^0 = $ a constant and its value is known.

The simplest situation is the assumption $\Delta C_P^0 = 0$. Then $\Delta H^0 = \Delta D = $ constant, and Eq. (6.53) is applicable. If $\Delta C_P^0 \approx \Delta a$, then

$$\Delta H^0 = \Delta A \cdot T + \Delta D \tag{6.59}$$

Combining Eq. (6.47) with Eq. (6.59), we get

$$\frac{d\,(\ln p_A^0)_{eq}}{dT} = \frac{\Delta A \cdot T + \Delta D}{RT^2} = \frac{1}{R}\left(\frac{\Delta A}{T} + \frac{\Delta D}{T^2}\right) \tag{6.60}$$

Integrating Eq. (6.60), we obtain

$$\ln\,(p_A^0)_{eq} = \frac{1}{R}\left(\Delta A \cdot \ln T - \frac{\Delta D}{T}\right) + \text{constant} = \frac{m}{T} + n \ln T + I \tag{6.61}$$

where m, n and I are lumped constants obtained from empirical C_P, ΔH and vapour pressure data. *Equation (6.61) is the most widely employed equation.* Since we are dealing with equilibrium vapour pressure, the subscript "eq" is generally omitted. Figure 6.3 shows vapour pressure vs. temperature curves for some liquid metals. The unit of vapour pressure is atm.

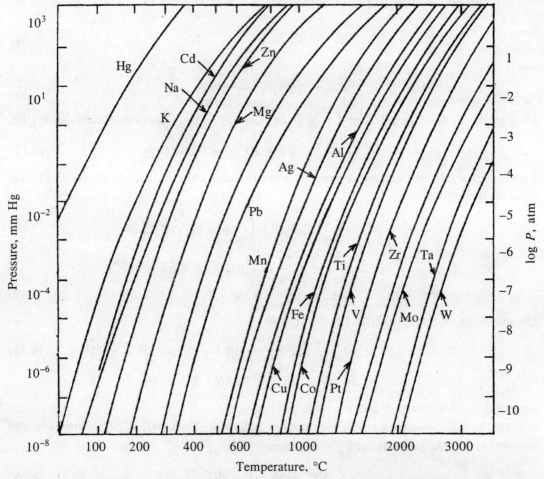

Fig. 6.3 Equilibrium vapour pressures of some metals as functions of temperature. (J.F. Elliott and M. Gleiser, *Thermochemistry for Steelmaking*, Vol. 1, Addison-Wesley (Mass.), 1960, p. 209.)

It may be mentioned that, if all terms in Eq. (6.58) are retained, then we obtain the most general vapour pressure vs. temperature relation as

$$\ln p_A^0 = \frac{m}{T} + n \ln T + lT + \frac{k}{T^2} + l \tag{6.62}$$

where l and k are two additional empirical constants. However, *this equation is rarely employed.*

6.4.5 Solid-Liquid-Vapour Equilibria: Triple point and metastability

Figure 6.2 has presented phase diagram for H_2O. There, the point 'O' is the common intersection of the lines for all three equilibria: (i) solid-vapour (SV), (ii) solid-liquid (SL), and (iii) liquid-vapour (LV). *O is known as the triple point.* Figure 6.4 presents the phase diagram for zinc. Both in Figs. 6.2 and 6.4, pressure means vapour pressure for SV and LV equilibria which involve the vapour. *For solid-liquid equilibria, pressure means total pressure P above O.* At triple point, it is vapour pressure for SL also. The existence of a unique triple point for unary systems can be predicted on the basis of *Gibbs Phase Rule*, which will be discussed in Chapter 10.

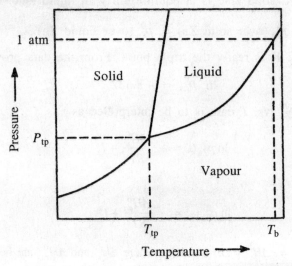

Fig. 6.4 Phase diagram for zinc (schematic).

Quite often, phase transformations occur very slowly. There are situations where thermodynamically stable states (i.e. ultimate equilibrium) are not attained even in centuries or even in long geologic periods. These are called *metastable phases*. They have stability, but not the ultimate stability. Some relevant examples are now given:

1. Cementite (Fe_3C) is most common in microstructures of iron-carbon alloys. However, cementite is metastable, and graphite is the stablest state of carbon.

2. Glass is an amorphous supercooled liquid. The crystalline state is the stablest. But glass can remain as such for ages at room temperature. Only upon heating for a prolonged period above a critical temperature does it crystallize.

3. A mixture of hydrogen and oxygen can be kept at room temperature for a very very long period without reaction. However, it is metastable. If it is ignited by a match stick, H_2 and O_2 react with explosive speed to form H_2O.

Thermodynamics is capable of quantitative treatment of metastable phases just like stable phases. In Fig. 6.2, for example, the dashed lines are extensions of solid lines, and represent metastable equilibria.

EXAMPLE 6.4   The vapour pressure of liquid zinc above its melting point (692 K) is given by the approximate formula:

$$\ln p_{Zn}^0 = \frac{-14.765 \times 10^3}{T} + 12.805$$

The heat of fusion of zinc is equal to 6673 J mol^{-1}. Derive a formula for the vapour pressure over solid zinc below 692 K.

Solution   At 692 K, solid zinc is at equilibrium with liquid zinc. Hence, at 692 K,

$$p_{Zn}^0 \text{ (over solid Zn)} = p_{Zn}^0 \text{ (over liquid Zn)}$$

In other words, 692 K is really the triple point. From the data provided, at 692 K,

$$\ln p_{Zn}^0 = -8.532 \qquad\qquad (E.6.3)$$

From Eq. (6.52), the p_{Zn}^0 vs. T data is to be interpreted as

$$\ln p_{Zn}^0(l) = -\frac{\Delta H_v^0}{RT} + I \qquad\qquad (E.6.4)$$

Also,

$$\ln p_{Zn}^0(s) = -\frac{\Delta H_s^0}{RT} + I' \qquad\qquad (E.6.5)$$

Now, heat of sublimation $= \Delta H_s^0 = \Delta H_v^0 + \Delta H_m^{0^*}$, where ΔH_v^0 and ΔH_m^0 are heats of vaporization and fusion, respectively. Hence,

$$\frac{\Delta H_s^0}{R} = 14.765 \times 10^3 + \frac{6673}{8.314} = 15.568 \times 10^3$$

Equating relations (E.6.3) and (E.6.5) at $T = 692$, we get

$$-8.532 = -\frac{15.568 \times 10^3}{692} + I'$$

i.e.

Thus,

$$l' = 13.965$$

$$\ln p_{Zn}^0 (s) = - \frac{15.568 \times 10^3}{T} + 13.965$$

6.5 Summary

1. For an isothermal process (reversible or irreversible) at temperature T,

   (a) $\Delta G = G_2 - G_1 = (H_2 - TS_2) - (H_1 - TS_1) = \Delta H - T \cdot \Delta S$, where 2, 1 designate state 2 and state 1, respectively.

   (b) For change of pressure,

   $$dG = RTd \, (\ln P), \quad \text{for ideal gas}$$

   $$= RTd \, (\ln f), \quad \text{for nonideal gas, and condensed substances}$$

   where f is known as fugacity. A gas tends to be more and more ideal, as pressure is lowered. Then, $f \to P$. At room temperature and above, and pressure not too high (say, lower than 10 atm), the common permanent gases may be taken as ideal.

2. The two alternative forms of Gibbs–Helmholtz (G–H) equation may be stated as:

   $$\left[\frac{\delta(\Delta G/T)}{\delta T} \right]_P = - \frac{\Delta H}{T^2} \quad \text{or} \quad \left[\frac{\delta(\Delta G/T)}{\delta(1/T)} \right]_P = \Delta H$$

   where ΔG, ΔH are changes of G and H, respectively, during isothermal processes.

3. Change of equilibrium phase transformation temperature of a pure substance in condensed state, due to change of pressure, is given by Clapeyron equation, viz.

   $$\left(\frac{dT}{dP} \right)_{eq} = \frac{\Delta V_{tr} \cdot T_{tr}}{\Delta H_{tr}}$$

   where ΔV_{tr} and ΔH_{tr} are changes in molar volume and enthalpy, respectively due to transformation.

4. For phase equilibria involving a pure condensed phase and its vapour, the Clausius–Clapeyron (CC) equation is applicable. The two forms of CC equation are

   $$\frac{d(\ln p_A^0)_{eq}}{dT} = \frac{\Delta H^0}{RT^2}, \quad \frac{d(\ln p_A^0)_{eq}}{d(1/T)} = - \frac{\Delta H^0}{R}$$

   where $(p_A^0)_{eq}$ is equilibrium (i.e. saturated) vapour pressure of substance A at temperature T, and ΔH^0 is change of molar enthalpy due to vapour formation.

5. The commonly used integrated form of CC equation is

$$\ln (p_A^0)_{eq} = \frac{m}{T} + n \ln T + I$$

where m, n, I are empirical constants.

6. The temperature at which solid, liquid and vapour of a pure substance co-exist at equilibrium, is invariant and is known as the *triple point*.

PROBLEMS

6.1 Find the change in free energy associated with isothermal expansion of 1 mole of an ideal gas from 2 atm to 1 atm pressure at 300 K.

6.2 Vapour pressure of water at 25°C is 23.76 mm Hg. Calculate the free energy change for transfer of 1 mole of water to vapour at 1 atm pressure at 25°C.

6.3 The variation of the volume of a liquid with pressure is given by: $V = V_0(1 - \beta P)$, where β is the compressibility coefficient and V_0 is the volume at low (i.e. virtually zero) pressure. Derive an expression of ΔG for a liquid due to change of pressure from P_1 to P_2. If, for water at 25°C, $V = 1.0029$ cm^3 g^{-1} at 1 atm, and $\beta = 49 \times 10^{-6}$ atm^{-1}, calculate ΔG per mole of water due to compression from 1 to 10 atm at 25°C. Would it make appreciable difference if β is assumed as zero?

6.4 Solid bismuth has a density of 9.80 g cm^{-3} at 25°C. Its coefficient of linear thermal expansion is 14.6×10^{-6}/°C. Density of liquid bismuth is 10.07 g cm^{-3}. The atomic mass of Bi is 209. What would be the change in melting temperature at a pressure of 1000 atm?

6.5 Calculate the minimum hydrostatic pressure required to stabilize Fe$_3$C at 298 K. Given:

(a) 3Fe(s) + C(graphite) = Fe$_3$C(s), $\Delta G^0 = 19,045$ J mol^{-1} at 298 K.
(b) The molar volumes of Fe$_3$C, graphite and Fe are 24.24, 5.34 and 7.10 cm$^3 \cdot$ mol^{-1}, respectively.
(c) Assume the molar volumes to be independent of pressure.

6.6 From the vapour pressure vs. temperature relationship for liquid silver, calculate (a) the heat of vaporization of liquid silver at the normal boiling temperature of 2147 K, and (b) heat capacity difference between liquid and gaseous silver.

6.7 Below the triple point (−56.2°C), the vapour pressure of solid CO$_2$ is given as

$$\ln p\,(\text{atm}) = -\frac{3116}{T} + 16.01$$

The molar heat of melting of CO$_2$ is 8330 joules. Calculate the vapour pressure of liquid CO$_2$ at 25°C, and explain why solid CO$_2$ evaporates rather than melts in an open container.

6.8 The chemical reaction inside the retort for extraction of zinc at high temperature, above boiling point of Zn, may be written as

$$ZnO(s) + C(s) = Zn(g) + CO(g)$$

The gaseous products of reaction enter a condenser (for condensing Zn(g) into liquid Zn) at 950°C and leave the condenser at 450°C and 1 atm. Assuming that the Zn vapour coming out of the condenser is at equilibrium with liquid Zn at 450°C, calculate the efficiency of recovery of Zn in the condenser.

Chapter 7

Activity, Equilibrium Constant and Standard Free Energy of Reactions

7.1 Introduction

7.1.1 Multicomponent Systems—Solutions and mixtures

The basic equations of thermodynamics, as derived and presented earlier, are general in nature. However, for the sake of convenience and systematic elucidation, we have restricted applications to: (a) physical processes, and (b) physico-chemical process of phase transformation in the one-component system.

A multicomponent system consists of more than one element or compound. Such a system may be classified as (a) mixture of pure elements and compounds and (b) solution. These are briefly described now.

Mixture of pure elements and compounds

An example of this is pure metal, pure oxygen gas and pure metal oxide contained in a closed vessel. It has three components. But each of them is present as pure, separate phases. In other words, it is a three-phase system consisting of mixture of pure elements and compound. Thus, a mixture is a heterogeneous system by definition.

Solution

A solution is a phase containing more than one component, where the components are mixed (i.e. dispersed) at atomic/molecular level. Aqueous solution with dissolved salts, acids etc. is an example. A NaCl solution in water has two components: NaCl, H_2O. It is a *binary solution*. If the solution contains H_2O, NaCl and KNO_3, then it is a *ternary solution*, and so on.

96

A solution has the following characteristics:

(i) Its composition can be changed (i.e. variable composition)

(ii) It is a single-phase substance (also known as a *homogeneous substance*).

In reality, we generally deal with a mixture of different solutions, which is a heterogeneous system consisting of more than one phase.

Some examples of such systems are:

- Aqueous solution and oil since they stay separately, there are two phases, but each of them is a solution.
- Mercury amalgam and aqueous solution co-existing as separate phases, but each of them is a solution.
- Air and aqueous solution, air being a solution of O_2, N_2, etc.
- Metallic solution, oxide solution and gas mixture co-existing as separate phases (3-phase system).

7.1.2 Physical, Chemical and Physico-chemical Processes

In a physical process, such as expansion/contraction of a gas, the composition of the substance remains the same. The system may be homogeneous or heterogeneous, substances may be pure or solutions. It does not matter. Here, the entire substance behaves uniformly.

In contrast, in chemical reactions, and physico-chemical processes such as phase transformation, diffusion, behaviours of individual components would be different, if it is a multicomponent system.

Some examples are as follows:

1. If an alloy of copper and gold is exposed to oxygen at high temperature, copper is preferentially oxidized to copper oxide. If HCl is added to an aqueous solution of NaCl and KOH, it will react only with KOH.

2. If a salt solution is cooled, water preferentially freezes out as pure ice. The aqueous solution, as a result, becomes more and more concentrated, and reaches solubility limit for salt, which then starts precipitating out as solid.

3. Similarly, when a liquid silver-copper solution is cooled below 780°C, it forms two separate phases, i.e. one solid solution rich in silver (designated as *α-phase*), and another solid solution rich in copper (designated as *β-phase*).

Thus, for applying the laws of thermodynamics to chemical and physico-chemical processes, each component is to be considered separately since it behaves differently from another. For a system consisting of pure substance, it is straightforward and can be done on the basis of what we know already. This has been illustrated in Chapter 6. However, application to solutions requires additional information in the form of additional terms, concepts, equations and data.

7.2 Thermodynamics of Solutions—Some introductory definitions and concepts

7.2.1 Composition Parameters—Mole fraction

Amongst various composition parameters, the mole fraction has been adopted as the fundamental and generally used one in chemical thermodynamics for nonaqueous solutions. Components in a solution (e.g. H_2O, NaCl etc. in aqueous solution) would be designated by subscripts 1, 2, ..., i, j, ... The general symbol is i.

$$X_i = \frac{n_i}{n_1 + n_2 + ... + n_i + n_j + ...} = \frac{n_i}{\sum_i n_i} = \frac{n_i}{n_T} \qquad (7.1)$$

where n_i denotes the number of moles of component i in the solution, n_T is the total number of moles in the solution, and X_i *is the mole fraction of component i in the solution.* By definition,

$$\sum_i X_i = 1 \qquad (7.2)$$

A gas mixture is also a solution, and an ideal gas mixture obeys *Dalton's Law of partial pressures*, viz.

$$p_1 + p_2 + ... + p_i + p_j + ... = \sum_i p_i = P_T \qquad (7.3)$$

$$X_i = \frac{p_i}{P_T} \qquad (7.4)$$

where p_i is partial pressure of component i. As already stated, the unit of pressure is conventionally standard atmosphere (i.e. 760 mm Hg) in chemical thermodynamics.

7.2.2 Partial and Integral Molar Properties

Consider an *extensive property* of a solution and let it be designated by the general symbol Q. Then Q may refer to U, H, G, V, S, A etc. *For further discussion, let us restrict the symbol Q to mean only molar properties of a solution.* Then,

$$Q' = n_T Q \qquad (7.5)$$

where Q' = the value of the quantity Q for the entire solution and Q = integral molar value of Q in the solution. At constant P and T,

$$Q' = f(n_1, n_2, ... n_i, ...) \qquad (7.6)$$

Differentiating Eq. (7.6), we get

$$dQ' = \left(\frac{\delta Q'}{\delta n_1}\right)_{\substack{P, T, n_2, \ldots, n_i, \\ \text{except } n_1}} dn_1 + \left(\frac{\delta Q'}{\delta n_2}\right)_{\substack{n_1, \ldots, n_i, \\ \text{except } n_2}} dn_2 + \ldots + \left(\frac{\delta Q'}{\delta n_i}\right)_{\substack{n_1, n_2 \ldots, \\ \text{except } n_i}} dn_i + \ldots \quad (7.7)$$

Let

$$\left(\frac{\delta Q'}{\delta n_i}\right)_{P, T, n_1, \ldots, \text{except } n_i} = \bar{Q}_i \quad (7.8)$$

Then at constant P and T,

$$dQ' = \bar{Q}_1 \, dn_1 + \bar{Q}_2 \, dn_2 + \ldots \bar{Q}_i \, dn_i + \ldots = \sum_i \bar{Q}_i \, dn_i \quad (7.9)$$

Combining Eq. (7.9) with Eqs. (7.1) and (7.5), we obtain

$$dQ = \frac{dQ'}{n_T} = \frac{\sum_i \bar{Q}_i \, dn_i}{n_T} = \sum_i \bar{Q}_i \, dX_i \quad (7.10)$$

For Gibbs free energy, we may write

$$(dG')_{P,T} = \sum_i \bar{G}_i \, dn_i, \quad (dG)_{P,T} = \sum_i \bar{G}_i \, dX_i \quad (7.11)$$

$\bar{Q}_i$ *is called partial molar value of Q of component i in the solution.* It is a function of composition of the solution. Its physical significance can be explained as follows.

Imagine a very large quantity of a solution. Add 1 g-mole of pure i into it. Upon dissolution, the total quantity of the solution increases by 1 g-mole. As a result, the value of Q' also increases. But the composition of the solution does not change significantly by addition, due to its large quantity. In this case, if the final pressure and temperature after dissolution is the same as the initial one, then the increase of Q' is equal to $\bar{Q}_i$. Therefore, $\bar{Q}_i$ *is the value of Q per mole of i in the solution.* Again,

$$\bar{Q}_i^m = \bar{Q}_i - Q_i^0 \quad (7.12)$$

where $Q_i^0 = Q$ per mole for pure i, and $\bar{Q}_i^m$ = change of Q upon dissolution of 1 mole of pure i. From Eq. (7.12), for Gibbs free energy per g-mole of i,

$$\bar{G}_i^m = \bar{G}_i - G_i^0 \quad (7.13)$$

where $\bar{G}_i$ = partial molar free energy of i in the solution and $\bar{G}_i^m$ = *partial molar free energy of mixing of i in the solution.*

7.2.3 Fugacity and Activity of a Component in a Solution

From the above discussions, it is evident that $\bar{G}_i$ is nothing but G per mole of component i in a solution. Therefore, in analogy with Eq. (6.9), we may write

$$\bar{G}_i^m = \bar{G}_i - G_i^0 = RT \ln (f_i/f_i^0) \tag{7.14}$$

where

    f_i = *fugacity of component i in solution*
    f_i^0 = fugacity of pure i(i.e. at standard state of i)

Activity of component i in a solution (a_i) is defined as

$$a_i = f_i/f_i^0 \tag{7.15}$$

From Eqs. (7.14) and (7.15),

$$\bar{G}_i^m = \bar{G}_i - G_i^0 = RT \ln a_i \tag{7.16}$$

In an ideal gas mixture, $p_i = f_i$. Hence,

$$\bar{G}_i^m = RT \ln (p_i/p_i^0) \tag{7.17}$$

Since $p_i^0 = 1$ atm by convention, Eq. (7.17) becomes

$$\bar{G}_i^m = RT \ln p_i \tag{7.18}$$

Therefore, from Eqs. (7.16) and (7.18), *for ideal gas mixture,*

$$a_i = p_i \tag{7.19}$$

However, it is to be noted that a_i *is dimensionless and the unit of p_i is standard atmosphere. Therefore, a_i is only numerically equal to the value of p_i in atm, for ideal gases.*

    The validity of Eq. (7.18) has also been verified by experiments. A closed container was partitioned into two sections with a semi-permeable membrane, which allowed passage to oxygen only. One section was evacuated. Air at 1 atmosphere was filled into the other section. Then sufficient time was allowed for attainment of equilibrium by permeation of oxygen through the membrane. It was then found that p_{O_2} in $O_2 + N_2$ mixture was equal to pressure of pure oxygen (p_{O_2}) in the section, which was initially evacuated. *In other words, in an ideal gas mixture, it is the partial pressure, which constitutes the thermodynamic parameter, not the total pressure.*

7.3 Equilibrium Constant (K) For a Chemical Reaction

7.3.1 Relationship of K with Standard Free Energy of Reaction

Consider the following reaction at constant temperature and pressure:

$$l\text{L} + m\text{M} + \ldots = q\text{Q} + r\text{R} + \ldots \tag{7.20}$$

where L, M, ... and Q, R, ... are general symbols for reactants and products and l, m, ... and q, r, ... denote number of moles. Then,

$$\Delta G = (q\bar{G}_Q + r\bar{G}_R + ...) - (l\bar{G}_L + m\bar{G}_M + ...) \qquad (7.21)$$

$$\Delta G^0 = (qG_Q^0 + rG_R^0 + ...) - (l\bar{G}_L^0 + m\bar{G}_M^0 + ...) \qquad (7.22)$$

where ΔG = actual change of free energy as a result of the reaction and ΔG^0 = value of ΔG if reactants and products are at their respective standard states at the temperature T of reaction. From Eqs. (7.21) and (7.22),

$$\Delta G - \Delta G^0 = \left\lfloor q(\bar{G}_Q - G_Q^0) + r(\bar{G}_R - G_R^0) + ... \right\rfloor - \left\lfloor l(\bar{G}_L - G_L^0) + m(\bar{G}_m - G_m^0) + ... \right\rfloor \qquad (7.23)$$

Combining with Eq. (7.16), Eq. (7.23) may be rewritten as

$$\Delta G - \Delta G^0 = RT[(q \ln a_Q + r \ln a_R + ...) - (l \ln a_L + m \ln a_M + ...)]$$

$$= RT \ln \left[\frac{a_Q^q \cdot a_R^r ...}{a_L^l \cdot a_M^m ...} \right] = RT \ln J \qquad (7.24)$$

The parameter J is known as activity quotient. At equilibrium,

$$(\Delta G)_{P,T} = 0 \qquad (5.44)$$

and hence,

$$\Delta G^0 = - RT \ln [J]_{\text{at eq}} = - RT \ln K \qquad (7.25)$$

where K is the *equilibrium constant* for reaction (7.20).

Again, combining Eqs. (7.24) and (7.25), we get

$$\Delta G = RT \ln (J/K) \qquad (7.26)$$

7.3.2 Comments on Application

Equation (7.25) constitutes the basis for equilibrium calculations at constant T and P. Equation (7.24) or (7.26) is employed for predicting feasibility of a reaction.

For an *irreversible (i.e. spontaneous) process,*

$$(\Delta G)_{T,P} < 0 \qquad (5.62)$$

This equation is satisfied only when

$$J/K < 1 \qquad (7.27)$$

It should be emphasized again that, by constant T and P in Eqs. (5.44) and (5.62), we mean that the initial and final temperature and pressure are the same. During the course of reaction, both T and P may vary in any manner without affecting conclusions.

Since the standard state of a substance is at 1 atm pressure, ΔG^0 of a reaction is function of temperature only. Therefore, the equilibrium constant K is also a function of temperature only.

Free energy is a state property. Hence, Hess' Law (see Section 3.3.4) is applicable. Thus,

$$\Delta G^0 = \text{(for reaction 7.20)} = \sum_{\text{products}} \Delta G_f^0 - \sum_{\text{reactants}} \Delta G_f^0$$

$$= \left\lfloor q\Delta G_f^0(Q) + r\Delta G_f^0(R) + ... \right\rfloor - \left\lfloor l\Delta G_f^0(L) + m\Delta G_f^0(M) + ... \right\rfloor \quad (7.28)$$

where ΔG_f^0 is *standard Gibbs free energy of formation of a compound* from its elements at temperature T. Obviously, ΔG_f^0 for an element is zero at any temperature, by definition. Therefore, if we have values of ΔG_f^0s, we can calculate ΔG^0 of a reaction. Hence, we shall concentrate on variation of ΔG_f^0 with T in the following sections, and also present a graphical method of representation of the same, known as the *Ellingham diagrams*.

7.4   Variation of ΔG^0 and K with Temperature

7.4.1   Derivation of Relationship of ΔG_f^0 with Temperature

On the basis of Gibbs–Helmholtz equation,

$$\left[\frac{\delta(\Delta G/T)}{\delta T}\right]_P = -\frac{\Delta H}{T^2} \quad (6.28)$$

we may write, for formation reactions at constant pressure,

$$\frac{d(\Delta G_f^0/T)}{dT} = -\frac{\Delta H_f^0}{T^2} \quad (7.29)$$

where ΔH_f^0 = standard enthalpy change for formation of a compound from its elements at temperature T. Since pressure is constant, Eq. (7.29) has been simplified and written as exact differential equation.

If there is no phase change, then on the basis of Kirchhoff's Law, we may write

$$\Delta H_f^0(T) = \Delta H_f^0(298) + \int_{298}^{T} \Delta C_p^0 \, dT \quad (7.30)$$

where

$$\Delta C_P^0 = (C_P^0)_{\text{compound}} - \sum_{\text{elements}} C_P^0 \quad (7.31)$$

By Eq. (6.57), we have

$$\Delta C_P^0 = \Delta a + \Delta b \cdot T + \Delta c \cdot T^{-2} \quad (7.32)$$

Combining Eqs. (7.29) to (7.32), we get

$$\frac{d\left(\dfrac{\Delta G_f^0}{T}\right)}{dT} = -\,\frac{\Delta H_f^0(298) + \displaystyle\int_{298}^{T}\left(\Delta a + \Delta b \cdot T + \dfrac{\Delta c}{T^2}\right)dT}{T^2}$$

$$= -\,\frac{1}{T^2}\left[\Delta H_f^0(298) + \Delta a(T - 298)\right.$$

$$\left. +\,\frac{\Delta b}{2}\,(T^2 - 298^2) - \Delta c\left(\frac{1}{T} - \frac{1}{298}\right)\right] \qquad (7.33)$$

Rearranging terms, we obtain

$$\frac{d\left(\dfrac{\Delta G_f^0}{T}\right)}{dT} = -\,\frac{1}{T^2}\left[\Delta H_f^0(298) - \Delta a \cdot 298 - \frac{\Delta b}{2}\cdot 298^2 + \Delta c \cdot \frac{1}{298}\right]$$

$$-\,\frac{\Delta a}{T} - \frac{\Delta b}{2} + \frac{\Delta c}{T^3} = -\,\frac{k}{T^2} - \frac{\Delta a}{T} - \frac{\Delta b}{2} + \frac{\Delta c}{T^3} \qquad (7.34)$$

where k is a constant. This is because $\Delta H_f^0(298)$ is also a constant. Integrating Eq. (7.34), we get

$$\frac{\Delta G_f^0}{T} = \frac{k}{T} - \Delta a \ln T - \frac{\Delta b}{2}T - \frac{\Delta c}{2T^2} + e \qquad (7.35)$$

where e is integration constant.

Arrangements of terms and generalization yield an equation of the following form:

$$\Delta G_f^0 = I + mT + gT \ln T + nT^2 + dT^{-1} \qquad (7.36)$$

where I, m, n, d, g are constants. They are lumped parameters based on the values of $\Delta H_f^0(298)$, Δa, Δb and Δc.

Equation (7.36) is the most general form of the variation of ΔG_f^0 with temperature. However, it is not normally employed since not all terms are significant. Terms which are small really lie within the error limits of experimental data on ΔG_f^0 or ΔH_f^0. Ignoring the terms $\Delta b \cdot T$ and $\Delta c \cdot T^{-2}$ in Eq. (7.32), since they have been mostly found to be small, we have

$$\frac{d\left(\dfrac{\Delta G_f^0}{T}\right)}{dT} = -\,\frac{1}{T^2}\left[\Delta H_f^0(298) + \Delta a \cdot T - \Delta a \cdot 298\right] \qquad (7.37)$$

Following the derivation procedure of Eq. (7.36), we may write

$$\Delta G_f^0 = I + mT + gT \ln T \tag{7.38}$$

Most thermochemical data books express temperature dependence of ΔG_f^0 in the form of Eq. (7.38). For many reactions, ΔH_f^0 *is approximately independent of temperature* (i.e. $\Delta C_P^0 \approx 0$) over a limited temperature range. Then Eq. (7.38) gets further simplified into

$$\Delta G_f^0 = I + mT \tag{7.39}$$

7.4.2 Alternative Derivation Procedure of ΔG_f^0 vs. T Relation

For an isothermal process, from Chapter 6,

$$\Delta G = \Delta H - T\Delta S \tag{6.3}$$

Therefore,

$$\Delta G_f^0(T) = \Delta H_f^0(T) - T\Delta S_f^0(T) \tag{7.40}$$

Noting the expressions for the variation of entropy with temperature from Section 4.3.2, if there is no phase change, then for a pure substance,

$$S_T^0 = S_{298}^0 + \int_{298}^{T} \frac{C_p^0}{T} dT \tag{7.41}$$

On the basis of the above, we may write

$$\Delta S_f^0(T) = \Delta S_f^0(298) + \int_{298}^{T} \frac{\Delta C_p^0}{T} dT \tag{7.42}$$

Hence, Eq. (7.40) becomes

$$\Delta G_f^0(T) = \left[\Delta H_f^0(298) + \int_{298}^{T} \Delta C_p^0 \, dT \right] - T\left[\Delta S_f^0(298) + \int_{298}^{T} \frac{\Delta C_p^0}{T} dT \right] \tag{7.43}$$

Combining Eq. (7.43) with Eq. (7.32), we can arrive at Eq. (7.36) as well. This is the alternative procedure for derivation of Eq. (7.36). If $\Delta C_p^0 = 0$, then both ΔH_f^0 and ΔS_f^0 are independent of temperature, and Eq. (7.39) is obtained. Then, combining Eqs. (7.39) and (7.40), we get

$$\Delta G_f^0 = \Delta H_f^0 - T \cdot \Delta S_f^0 = I + mT \tag{7.44}$$

EXAMPLE 7.1 Calculate ΔG for the process: Al(s) = Al(l) at 1000 K.

Given: for aluminium, melting point = 933 K; heat of fusion at melting point = 10.9 $\times 10^3$ J mol^{-1}; C_P = 32.5 and 29.3 J mol^{-1}K^{-1} for solid and liquid, respectively.

Solution

$$\Delta G_{1000} = \Delta H_{1000} - 1000(\Delta S)_{1000} \qquad (E.7.1)$$

$$\Delta H_{1000} = \Delta H_{933} + \int_{933}^{1000} (\Delta C_P)\, dT$$

$$= 10.9 \times 10^3 + (29.3 - 32.5)(1000 - 933)$$

$$= 10,686 \text{ J mol}^{-1}$$

$$\Delta S_{1000} = \Delta S_{933} + \int_{933}^{1000} \frac{\Delta C_P}{T}\, dT$$

$$= \frac{10.9 \times 10^3}{933} + (29.3 - 32.5) \ln \frac{1000}{933}$$

$$= 11.461 \text{ J mol}^{-1} \text{ K}^{-1}$$

Hence, from Eq. (E.7.1), we have

$$\Delta G_{1000} = 10,686 - 1000 \times 11.461 = -775 \text{ J mol}^{-1}$$

EXAMPLE 7.2 (i) Calculate ΔG^0 at 600 K for one mole of the reaction:

$$NiO\ (s) + Co\ (s) = CoO\ (s) + Ni\ (s)$$

(ii) How much error is involved if ΔC_P for this reaction is assumed to be zero?

Given: (i) ΔH_f^0 at 298 K for CoO and NiO are: −239 and −244.6 kJ mol^{-1}, respectively.

(ii) S_{298}^0 values are 30.0, 52.9, 29.8, 38.1 J mol^{-1}K^{-1} for Co, CoO, Ni, NiO, respectively.

(iii) The values of C_P (J mol^{-1}K^{-1}) are as follows:

Co: $21.4 + 14.3 \times 10^{-3}\ T - 0.88 \times 10^5\ T^{-2}$

CoO: $48.28 + 8.54 \times 10^{-3}\ T + 1.7 \times 10^5\ T^{-2}$

Ni: $32.6 - 1.97 \times 10^{-3}\ T - 5.59 \times 10^5\ T^{-2}$

NiO: $-20.9 + 157.2 \times 10^{-3}\ T + 16.3 \times 10^5\ T^{-2}$

Solution (i) From Eq. (6.3),

$$\Delta G_{600}^0 = \Delta H_{600}^0 - 600 \times \Delta S_{600}^0 \qquad (E.7.2)$$

From Eq. (7.30),

$$\Delta H_{600}^0 = \Delta H_{298}^0 + \int_{298}^{600} \Delta C_P\, dT \qquad (E.7.3)$$

From Eq. (7.42),

$$\Delta S_{600}^0 = \Delta S_{298}^0 + \int_{298}^{600} \frac{\Delta C_P}{T}\, dT \qquad (E.7.4)$$

[*Note:* If there are phase transformations, then the more general form of Kirchhoff equation, as given in Chapters 3 and 4, are to be employed in Eqs. (E.7.3) and (E.7.4).]

Now,

$$\Delta H^0_{298} = \Delta H^0_{f,\,298}(CoO) - \Delta H^0_{f,\,298}(NiO), \text{ (from Hess' Law)}$$

$$= (-239 + 244.6) \times 10^3 = 5600 \text{ J mol}^{-1}$$

$$\Delta S^0_{298} = \left\lfloor S^0_{298}(CoO) + S^0_{298}(Ni) \right\rfloor - \left\lfloor S^0_{298}(NiO) + S^0_{298}(Co) \right\rfloor$$

$$= 52.9 + 29.8 - 38.1 - 30.0 = 14.6 \text{ J mol}^{-1}K^{-1}$$

$$\Delta C_P = [C_P(CoO) + C_P(Ni)] - [C_P(NiO) + C_P(Co)] \tag{E.7.5}$$

From the data provided and from Eq. (E.7.5),

$$\Delta C_P = 80.38 - 164.9 \times 10^{-3}T - 19.31 \times 10^5 T^{-2} \text{ J mol}^{-1}K^{-1}$$

Solving Eqs. (E.7.3) and (E.7.4), we get

$$\Delta H^0_{600} = 4250 \text{ J}, \quad \Delta S^0_{600} = 12.99 \text{ J K}^{-1}$$

$$\Delta G^0_{600} = 4250 - 600 \times 12.99 = -3544 \text{ joules}$$

(ii) If ΔC_P of the reaction is assumed to be zero, then

$$\Delta H^0_{600} = \Delta H^0_{298}, \quad \Delta S^0_{600} = \Delta S^0_{298}$$

Therefore,

$$\Delta G^0 = 5600 - 600 \times 14.6 = -3160 \text{ joules}$$

7.4.3 Variation of Equilibrium Constant with Temperature

On the basis of Gibbs–Helmholtz Equations (6.28) and (6.29), we may write

$$\frac{d(\Delta G^0/T)}{dT} = -\frac{\Delta H^0}{T^2} \tag{7.45}$$

$$\frac{d(\Delta G^0/T)}{d(1/T)} = \Delta H^0 \tag{7.46}$$

Combining these with the relation: $\Delta G^0 = -RT \ln K$, the variation of K with T is given by either

$$\frac{d(\ln K)}{dT} = \frac{\Delta H^0}{RT^2} \tag{7.47}$$

or

$$\frac{d(\ln K)}{d(1/T)} = -\frac{\Delta H^0}{R} \tag{7.48}$$

The above are alternative forms of the *Van't Hoff equation.*

7.5 Ellingham Diagrams

Ellingham diagrams are basically graphical representations (i.e. nomograms) of ΔG^0 vs. T relations for chemical reactions. They have some more features, which will be explained in Chapter 8. They are so called after C.J.T. Ellingham who first constructed them in 1944.

Most common Ellingham diagrams are concerned with free energies of formation of compounds like oxides, sulphides, and chlorides. Figure 7.1 presents the Ellingham diagram for formation of oxides, and is the most common and widely used diagram in metallurgy and materials science.

7.5.1 Basic Features

(i) The formation reactions for oxides in Fig. 7.1 may be generalized as

$$\frac{2x}{y} M + O_2(g) = \frac{2}{y} M_x O_y \tag{7.49}$$

where the values of x and y will depend on the specific compound. M and $M_x O_y$ are general symbols for metal and metal oxide, respectively. They can be solid, liquid or gaseous, depending on temperature. Specific examples are:

$$2Ni + O_2(g) = 2NiO \tag{7.50}$$

$$Si + O_2(g) = SiO_2 \tag{7.51}$$

$$\frac{4}{3}Al + O_2(g) = \frac{2}{3}Al_2O_3 \tag{7.52}$$

Free energy is an extensive property. Hence, the value of ΔG of a reaction would depend on the number of moles involved. In the Ellingham diagram, 1 mole of oxygen constitutes the basis in the oxide formation reaction. Thus, *the values of ΔG_f^0 are per mole of O_2.*

(ii) The ΔG_f^0 vs. T curve for a reaction consists of some interconnected straight lines. As long as there is no phase change either in the element or its oxide, ΔG_f^0 is assumed to vary linearly with temperature in accordance with Eq. (7.44). The slope of a line changes if a phase change occurs in the element or the oxide. M, B denote, respectively melting and boiling points of the element. For the oxides, these are denoted by $[M]$, $[B]$.

Let us take the example of the formation of ZnO, as shown in Fig. 7.1. Liquid zinc boils at 907°C i.e. 1180 K (at point B), where the slope of the line changes upwards. At point B, liquid Zn and gaseous Zn at 1 atm pressure co-exist at equilibrium, and from the equilibrium criteria, i.e. Eqs. (5.44) and (6.31), for

$$Zn(l) = Zn(g, 1atm), \qquad \Delta G^0 = G_{Zn}^0(g) - G_{Zn}^0(l) = 0 \tag{7.53}$$

Hence, at point B, ΔG^0 for the reaction

$$2\,Zn + O_2(g) = 2\,ZnO(s) \tag{7.54}$$

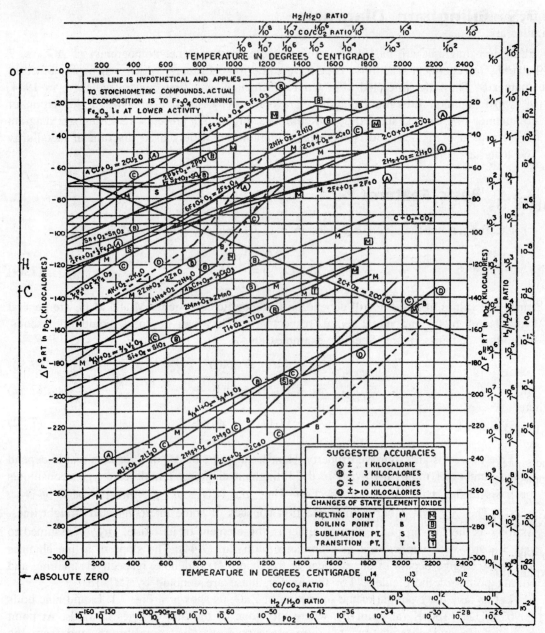

Fig. 7.1 Ellingham diagram giving standard free energies for formation of oxides from their respective elements, as function of temperature [F.D. Richardson and J.H.E. Jeffes, *J. Iron Steel Inst.*, **160**, 261 (1948)].

is the same for both Zn(l) and Zn(g). *Thus, there is no discontinuity in the curve at B.* With similar arguments, there will be no discontinuity at points M, $[M]$, $[B]$.

(iii) The changes in slopes of lines at M, B etc. can be explained as follows. From Eq. (7.44), the slope of the line of ΔG_f^0 vs. T is $-\Delta S_f^0$ and the intercept of the line at absolute zero (i.e. 0 K) is ΔH_f^0. For ZnO formation below 1180 K, let $\Delta S_f^0 = \Delta S_1^0$ and $\Delta H_f^0 = \Delta H_1^0$. Then,

$$\Delta S_1^0 = 2 S_{ZnO(s)}^0 - 2 S_{Zn(l)}^0 - S_{O_2(g)}^0 \tag{7.55}$$

$$\Delta H_1^0 = 2 H_{ZnO(s)}^0 - 2 H_{Zn(l)}^0 - H_{O_2(g)}^0 \tag{7.56}$$

Above 1180 K, let $\Delta S_f^0 = \Delta S_2^0$ and $\Delta H_f^0 = \Delta H_2^0$. Then,

$$\Delta S_2^0 = 2 S_{ZnO(s)}^0 - 2 S_{Zn(g)}^0 - S_{O_2(g)}^0 \tag{7.57}$$

$$\Delta H_2^0 = 2 H_{ZnO(s)}^0 - 2 H_{Zn(g)}^0 - H_{O_2(g)}^0 \tag{7.58}$$

Combining Eq. (7.55) with Eq. (7.57), and Eq. (7.56) with Eq. (7.58), we get

$$\Delta S_2^0 - \Delta S_1^0 = 2 \left[S_{Zn(l)}^0 - S_{Zn(g)}^0 \right] = -2 \Delta S_v^0 (Zn) \tag{7.59}$$

$$\Delta H_2^0 - \Delta H_1^0 = 2 \left[H_{Zn(l)}^0 - H_{Zn(g)}^0 \right] = -2 \Delta H_v^0 (Zn) \tag{7.60}$$

where ΔS_v^0 and ΔH_v^0 are *entropy and enthalpy of vaporization of liquid zinc.* Since ΔS_v^0 is a positive quantity, $\Delta S_2^0 - \Delta S_1^0 < 0$, i.e. $\Delta S_2^0 < \Delta S_1^0$. As, the slope of ΔG_f^0 vs. T curve is $-\Delta S_f^0$ and $-\Delta S_2^0 > -\Delta S_1^0$, the change of slope is positive at point B. Similarly, at point M, the slopes of lines change in the upward direction, whereas for points $[M]$ and $[B]$, they are opposite.

7.5.2 Nature of Slopes

In Fig. 7.1, all lines slope upwards except those for the formation of CO and CO_2. The explanations for this are as follows: For formation of a metal oxide such as

$$2M \ (s, l) + O_2 (g) = 2 \, MO \ (s, l) \tag{7.61}$$

$$\Delta S_f^0 = 2 S_{MO(s, l)}^0 - 2 S_{M(s, l)}^0 - S_{O_2(g)}^0 \tag{7.62}$$

Now, entropy is a measure of disorder of a substance (see Section 4.4.2). More the disorder, the higher is its entropy. Since the gas is a highly disordered state in comparison to solid or liquid, entropy of oxygen gas is much larger than the entropy of metal and oxide in Eq. (7.62). Therefore, $\Delta S^0 \approx -S_{O_2(g)}^0$, and hence the slope $(= -\Delta S^0)$ is positive. For,

$$C \ (s) + O_2(g) = CO_2(g) \tag{7.63}$$

one mole of O_2 disappears, but 1 mole of gaseous CO_2 forms. Hence, ΔS^0 is very small and the line is horizontal in the Ellingham diagram. For,

$$2C(s) + O_2(g) = 2CO(g) \tag{7.64}$$

The situation is reverse of Eq. (7.62). One extra mole of gas is produced. Hence, ΔS^0 for the reaction of CO is positive and the slope is downwards.

The above features of CO and CO_2 have very important consequence in extraction and refining of metals as well as in chemical industries.

7.5.3 Assessment of Relative Stabilities of Compounds

Graphical representations are not precise. Hence, for getting the precise values of free energies of formation, thermochemical data tables or equations such as Eq. (7.38) should be used. The Ellingham diagram is for finding out approximate values of ΔG_f^0 quickly. In addition, it is a quick guide for prediction of relative stabilities of compounds. For example, at 1000°C, from Fig. 7.1, for the reaction

$$\text{Si (pure, s)} + O_2 \text{ (g, 1 atm)} = SiO_2 \text{ (pure, s)} \tag{7.65}$$

$$\Delta G_{65} = \Delta G_{65}^0 = G_{SiO_2}^0 - G_{Si}^0 - G_{O_2}^0 = -678 \text{ kJ per mole } O_2 \text{ (approximate)} \tag{7.66}$$

(after conversion of kcal into kJ)

Similarly, for reaction (7.64) and with C, CO, O_2 at their respective standard states, viz. pure solid carbon, CO and O_2 at 1 atm each,

$$\Delta G_{64} = \Delta G_{64}^0 = -448 \text{ kJ per mole } O_2 \text{ (approximate)}$$

As points of clarification, ΔG_{65}, ΔG_{64} *denote ΔG for reactions (7.64) and (7.65), respectively.* They are equal to ΔG_{65}^0 and ΔG_{64}^0 respectively, since all reactants and products are at their respective standard states at the temperature under consideration. That is why solids and liquids have to be pure and gases at 1 atm.

Subtracting reaction (7.65) from (7.64), we obtain

$$2C(s, \text{pure}) + SiO_2 (s, \text{pure}) = 2CO \text{ (g, 1 atm)} + \text{Si (pure, s)} \tag{7.67}$$

for which $\Delta G_{67}^0 = \Delta G_{64}^0 - \Delta G_{65}^0 = +230$ kJ per mole of reaction (7.67).

As discussed in Section 5.6, a positive value of ΔG (in this case, ΔG^0) means that reaction (7.67) is not feasible at constant temperature and pressure. Therefore, pure carbon cannot reduce pure SiO_2 at 1000°C and $p_{CO} = 1$ atm. In other words, SiO_2 is more stable as compared to CO at their respective standard states at 1000°C. This is what is meant by relative stabilities of compounds.

In the Ellingham diagram, the line for SiO_2 is below that of CO at 1000°C. To generalize, the line for a stabler oxide would be below that of a less stable oxide. Therefore, Al_2O_3 is stabler than SiO_2.

At 2000°C, Fig. 7.1 shows that CO is stabler than SiO_2. Hence, carbon can reduce SiO_2.

This interesting feature of the oxide system allows carbon to reduce any oxide into metal if we go to a sufficiently high temperature. This is the thermodynamic basis for extraction of metals from oxides, with carbon as reducing agent.

It is to be specially noted that the actual criterion for feasibility of a process/reaction is $(\Delta G)_{T, P} < 0$ [see Eq. (5.62)]. ΔG^0 is equal to ΔG only when all reactants and products are at their respective standard states, as illustrated above. More general criterion (i.e. ΔG) for all states, both non-standard and standard, will be taken up in subsequent chapters.

EXAMPLE 7.3 Predict the relative thermodynamic stabilities of solid Si_3N_4 and BN in a mixture of Si_3N_4 and B at 1 atm pressure. Assume that all components are pure. BN and Si_3N_4 are high temperature ceramic materials having technological applications (Johnson and Stratcher, Chapter 5).

Solution   Consider the reaction:

$$Si_3N_4(s) + 4B(s) = 4BN(s) + 3Si(s,l) \qquad (E.7.6)$$

Since all reactants and products are pure and under 1 atm pressure, ΔG for reaction (E.7.6) is the same as ΔG^0. Hence, if the reaction is at equilibrium at a temperature T, then

$$\Delta G_T^0 = 0 \qquad (E.7.7)$$

The forward reaction would occur if $\Delta G_T^0 < 0$, in which case, BN would have more stability than Si_3N_4. On the other hand, if $\Delta G_T^0 > 0$, then Si_3N_4 will be stabler than BN.

From Hess' Law, at temperature T,

$$\Delta G_T^0 = 4\,\Delta G_f^0(\text{BN}) - \Delta G_f^0(\text{Si}_3\text{N}_4) \qquad (E.7.8)$$

(i) Situation 1   Assume Si is solid.

$$B(s) + \frac{1}{2}N_2(g) = BN(s), \quad \Delta G^0 = -108,\, 800 + 40.6T \text{ J mol}^{-1} \quad \text{at 1200–2300 K.}$$

$$3Si(s) + 2N_2(g) = Si_3N_4(s), \quad \Delta G^0 = -753,\, 200 + 336.4T \text{ J mol}^{-1} \quad \text{at 298–1680 K.}$$

Substituting the above values into Eq. (E.7.8), the equilibrium temperature (T_{eq}) is obtained from the following equation:

$$0 = 4(-108,\, 800 + 40.6T_{eq}) - (-753,\, 800 + 336.4T_{eq})$$

yielding T_{eq} = 1831 K. However, Si melts at 1680 K, and this temperature is above that. Therefore, this is an invalid solution.

(ii) Situation 2   Assume Si to be a liquid.

$$3Si(l) + 2N_2(g) = Si_3N_4(s), \text{ for which } \Delta G^0 = -892,\, 950 + 419.3T \text{ J mol}^{-1} \text{ at 1680–1800 K.}$$

Combining this with ΔG_f^0 of BN, we get

$$0 = 4(-108,\, 800 + 40.6\,T_{eq}) - (-\,892,\, 950 + 419.3T_{eq}) = +457,\, 750 - 256.9T_{eq}$$

yielding $T_{eq} = 1782$ K. This is a valid solution. If $T > T_{eq}$, then $\Delta G_T^0 < 0$. Hence, BN would be relatively stable above 1782 K. Below 1782 K, Si_3N_4 would be relatively stable.

7.6 Summary

1. The free energy criterion is useful for application at constant pressure and temperature (see Chapter 5). Of these, the constant pressure restriction is not significant for condensed phases. But the process must be isothermal. Hence, from now on, constant temperature will be assumed in all derivations in this and subsequent chapters, unless otherwise stated.

2. In a solution, various components have been designated as 1, 2, ..., i, j, ..., the general symbol being i. Mole fraction is employed as the general composition parameter in thermodynamics of solution of non-aqueous systems.

$$\text{Mole fraction of component } i = X_i = \frac{n_i}{\sum_i n_i} = \frac{n_i}{n_T}$$

where n_i is the number of moles of i.

3. Some symbols and definitions of state properties for solution thermodynamics have been discussed. These are being summarized below with Gibbs free energy as example.

G' = value of G for the entire solution

$$G = \frac{G'}{n_T} = \text{integral molar free energy}$$

$$\overline{G}_i = \left(\frac{\delta G'}{\delta n_i} \right)_{P,T,n_1,n_2,\,...,\,\text{except } n_i} = \text{partial molar free energy of component } i$$

$$\overline{G}_i^m = \overline{G}_i - G_i^0 = \text{partial molar free energy of mixing of } i$$

where G_i^0 is free energy per mole of i in standard state

$$\Delta G^m = G - G^0 = \text{integral molar free energy of mixing}$$

4. $$\overline{G}_i^m = RT \ln \left(\frac{f_i}{f_i^0} \right) = RT \ln a_i$$

where f_i is fugacity and a_i is the activity of component i in a solution. For an ideal gas mixture, $a_i = p_i$, where p_i is partial pressure of i in atm.

5. For the chemical reaction

$$l\text{L} + m\text{M} + ... = q\text{Q} + r\text{R} + ..., \text{ at constant } T$$

we have

$$\Delta G = \Delta G^0 + RT \ln J, \text{ where } J = \text{activity quotient} = \frac{a_Q^q \times a_R^r \times \dots}{a_L^l \times a_M^m \times \dots}$$

At equilibrium at constant P and $\Delta G^0 = - RT \ln [J]_{eq} = - RT \ln K$, where K is the equilibrium constant.

6. According to Hess' Law,

$$\Delta G^0 = \underset{\text{products}}{\Sigma \Delta G_f^0} - \underset{\text{reactants}}{\Sigma \Delta G_f^0}$$

where ΔG_f^0 denotes free energy of formation of a compound from its elements.

7. ΔG_f^0 is a function of temperature and the most common form of the relation is:

$$\Delta G_f^0 = I + mT + gT \ln T$$

where I, m, g are empirical constants.

8. Ellingham diagrams are specially constructed nomograms, giving variation of ΔG^0 with temperature. Their properties and usefulness have been discussed by taking the example of formation of inorganic oxides from elements.

PROBLEMS

7.1 Calculate the equilibrium constant for the following reaction at 298 K:

$$C \text{ (graphite)} + 2H_2 \text{(g)} = CH_4 \text{(g)}$$

Use the following data for the reaction at 298 K:

$$\Delta H^0 = -74.85 \times 10^3 \text{ J mol}^{-1}, \qquad \Delta S^0 = - 80.25 \text{ J mol}^{-1}\text{K}^{-1}$$

7.2 Calculate the molar value of ΔG for Si(s) $\rightarrow$ Si(l) at 1500°C. The equilibrium melting point of solid silicon is 1410°C. Assume $\Delta C_P = 0$.

7.3 Calculate: (i) the entropy of transformation of Ti(α) to Ti(β) at the normal transformation temperature of 882°C, and (ii) ΔG of transformation at 930°C.

7.4 For 1 mole of the reaction: Ni(s) + 1/2 O$_2$(g) = NiO(s), calculate ΔH^0, ΔU^0, ΔS^0 and ΔG^0 at 500 K.

7.5 The standard free energy of formation of one mole of MgO is given as

$$\Delta G_f^0 = - 603.9 \times 10^3 - 5.355T \ln T + 393.1T \text{ joules}$$

Calculate the enthalpy and entropy of formation of MgO at 300 K.

7.6 Using the Ellingham diagram for oxides, show which metal will not get oxidized in superheated steam at 1000°C–Ni or Cr? The conclusions should be based on quantitative values of ΔG^0.

7.7 Predict the relative stability of Cr(s) and $Cr_{23} C_6$(s) at 1500 K.

   Given: for $Cr_2O_3(s) + \dfrac{12}{23} C(s) = \dfrac{2}{23} Cr_{23}C_6(s) + \dfrac{3}{2} O_2(g)$

$$\Delta G^0 = 1084.5 \times 10^3 - 263.17T \text{ J mol}^{-1}$$

7.8 For dissociation of silver oxide, i.e. $Ag_2O(s) = 2Ag(s) + \dfrac{1}{2} O_2(g)$, the equilibrium constant is 0.81 at 450 K. Calculate ΔH^0, ΔS^0 and ΔG^0 for the reaction at 300 K for one mole of the dissociation reaction.

7.9 The free energy of formation (ΔG_f^0) of NiO is given as −150.25 kJ/mol, and −133.43 kJ/mol at 1000 K and 1200 K, respectively. From these data, calculate
(a) equilibrium constant at each temperature
(b) ΔH_f^0 and ΔS_f^0 of NiO.

Equilibria Involving Ideal Gases and Pure Condensed Phases

8.1 Introduction

In Section 7.3, we have derived the basic equations for equilibrium calculations as well as prediction of feasibility for chemical reactions. For reaction (7.20),

$$J = \frac{a_Q^q \cdot a_R^r \cdots}{a_L^l \cdot a_M^m \cdots} \qquad (8.1)$$

Also,
$$K = [J]_{\text{at eq}} \qquad (8.2)$$

where a_i is activity of component i in a substance (pure or solution).

Activity is an abstract concept and has no ultimate utility to us. What we are concerned with are calculations of compositions of phases at chemical equilibrium. This requires knowledge of activity vs. composition relation for a solution. This topic will be taken up and elaborately discussed in Chapter 9. To sum up, generally speaking, activity vs. composition relations in solid and liquid solutions are based on experimental measurements, which constitute a major programme of thermodynamic measurements, as we shall see later.

However, we do not require experimental data on activity vs. composition relation for ideal gases and gas mixtures as well as pure condensed phases. Hence, in this chapter, we are taking up only equilibrium calculations involving these. Further reaction equilibrium calculations will be analyzed only in a later chapter after thermodynamics of solutions is presented.

115

8.2 Equilibrium Calculations for Reactions Involving Ideal Gases

In metallurgical and materials engineering, we are primarily concerned with systems and processes occurring at high temperatures. Pressures of gases are also generally not too high. As discussed in section 6.2.3, the gases and gas mixtures in the above situations may be treated as ideal without any significant errors. Further, as pointed out in section 7.2.3, in ideal gas mixture, $a_i = p_i$, where p_i is partial pressure of component i. It has also been derived in section 7.2.3 that, for thermodynamic calculations on the above basis, the unit of pressure is to be taken as atm, which is the symbol for standard atmosphere (1 standard atmosphere = 760 mm Hg). Moreover, for an ideal gas mixture,

$$X_i = \frac{p_i}{P_T} \tag{7.4}$$

where

X_i = mole fraction of component i in the mixture

= volume fraction of component i in the mixture,

P_T = total pressure of all components, i.e.

$$P_T = p_1 + p_2 + \dots + p_i + p_j + \dots = \sum_i p_i \tag{7.3}$$

In view of the above, we do not require any special knowledge of activity-composition relations for equilibrium calculations involving ideal gases. It is a straightforward application of equilibrium relations already derived in Section 7.3. In this chapter, calculation procedures shall be illustrated through a few examples.

It may also be noted that, at high temperatures (say, above 700–800°C), gaseous reactions are very fast and, hence, chemical equilibria are attained easily. Therefore, the actual composition of a gas mixture inside a high temperature furnace is generally assumed the same as that at equilibrium. This makes such equilibrium calculations a very useful tool for applications to real processes in industries and laboratories.

EXAMPLE 8.1 H_2-H_2O and CO-CO_2 gas mixtures are frequently employed in high temperature experiments in laboratory to obtain a gas phase with known partial pressure of oxygen (p_{O_2}). These gaseous compounds along with others are also common in metallurgical and materials processing as well as combustion of fuels at high temperature.

Solution For the reaction

$$H_2(g) + 1/2 O_2(g) = H_2O(g) \tag{E.8.1}$$

we have

$$\Delta G_1^0 = -239.534 \times 10^3 + 8.139T \ln T - 9.247T, \text{ J mol}^{-1} \tag{E.8.2}$$

(*Note:* mol^{-1} means per mole of reaction as written.)

Again, from Eq. (7.25),

$$\Delta G_1^0 = - RT \ln K_1 = - RT \ln \left[\frac{a_{H_2O}}{a_{H_2} \times a_{O_2}^{1/2}} \right]_{eq}$$

$$= - RT \ln \left[\frac{p_{H_2O}}{p_{H_2} \times p_{O_2}^{1/2}} \right]_{eq} \tag{E.8.3}$$

From Eqs. (E.8.2) and (E.8.3), and noting that $R = 8.316$ J mol^{-1}K^{-1}, we have

$$\ln \left[\frac{p_{H_2O}}{p_{H_2} \times p_{O_2}^{1/2}} \right]_{eq} = \frac{239.534 \times 10^3}{8.316T} - \frac{8.139 \ln T}{8.316} + \frac{9.247}{8.316} \tag{E.8.4}$$

As a sample calculation, at 2000 K, if we require $P_{O_2} = 10^{-10}$ atm, then from Eq. (E.8.4), the corresponding $\left(\frac{p_{H_2O}}{p_{H_2}} \right)$ ratio should be 3.4×10^{-2}.

Similarly, for the reaction

$$CO\,(g) + \frac{1}{2} O_2\,(g) = CO_2\,(g) \tag{E.8.5}$$

$$\Delta G_5^0 = -282.42 \times 10^3 + 86.82T,\ J\ mol^{-1}$$

$$= - RT \ln K_5 = - RT \ln \left[\frac{p_{CO_2}}{p_{CO} \times p_{O_2}^{1/2}} \right]_{eq} \tag{E.8.6}$$

Therefore,

$$\ln \left[\frac{p_{CO_2}}{p_{CO} \times p_{O_2}^{1/2}} \right]_{eq.} = \frac{1}{8.316} \left[\frac{282.42 \times 10^3}{T} - 86.82 \right] \tag{E.8.7}$$

EXAMPLE 8.2 A gas mixture consisting of 1 mole SO_2 and 0.6 mole O_2 is introduced into a furnace at 1000 K. The total pressure $(P_T) = 1$ atm. Calculate the composition after the mixture attains equilibrium in the furnace.

Solution For the reaction

$$SO_2\,(g) + 1/2\ O_2\,(g) = SO_3\,(g) \tag{E.8.8}$$

we have

$$\Delta G_8^0 = -94.558 \times 10^3 + 89.37T,\ J\ mol^{-1} \tag{E.8.9}$$

Thus, at 1000 K,

$$K_8 = 1.87 = \frac{p_{SO_3}}{p_{SO_2} \times p_{O_2}^{1/2}} \qquad \text{(E.8.10)}$$

Let y mole of SO_3 be present in the equilibrium gas mixture. Then, the gas would consist of

$$(1-y) \text{ mole } SO_2, \ (0.6 - 0.5y) \text{ mole } O_2, \ y \text{ mole } SO_3$$

Hence,

$$n_T = 1 - y + 0.6 - 0.5y + y = 1.6 - 0.5y \text{ mole} \qquad \text{(E.8.11)}$$

Combining Eq. (E.8.11) with Eqs. (7.1), (7.3) and (7.4), we get

$$p_{SO_2} = X_{SO_2} \cdot P_T = \frac{n_{SO_2}}{n_T} P_T = \frac{1 - y}{1.6 - 0.5y} \times 1 \qquad \text{(E.8.12)}$$

Similarly,

$$p_{O_2} = \frac{0.6 - 0.5y}{1.6 - 0.5y}, \qquad p_{SO_3} = \frac{y}{1.6 - 0.5y} \qquad \text{(E.8.13)}$$

From Eqs. (E.8.10), (E.8.12), and (E.8.13),

$$\frac{\left[\dfrac{y}{1.6 - 0.5y}\right]}{\left[\dfrac{1 - y}{1.6 - 0.5y}\right]\left[\dfrac{0.6 - 0.5y}{1.6 - 0.5y}\right]^{1/2}} = 1.87 \qquad \text{(E.8.14)}$$

Solving Eq. (E.8.14), $y = 0.49$ mole, and hence the equilibrium gas mixture will consist of

$$37.5 \text{ vol\% } SO_2, \qquad 26.5 \text{ vol\% } O_2 \text{ and } 36.0 \text{ vol\% } SO_3 \qquad \text{(Ans.)}$$

[*Note:* In a gas mixture, composition is expressed in volume per cent, which is $100 \times$ vol. fraction (i.e. $100 \times$ mole fraction as per Gas, Laws).]

EXAMPLE 8.3 A gas mixture consisting of 20% CO, 20% CO_2, 10% H_2, and 50% N_2 is fed into a furnace at 900°C. Find the equilibrium composition of the gas inside the furnace.

Solution As already mentioned, percentage composition in a gas mixture is by convention volume per cent, which is the same as mole percent.

At high temperature, chemical reaction would generate $H_2O(g)$ as seen from the following relation:

$$CO_2(g) + H_2(g) = CO(g) + H_2O(g) \qquad \text{(E.8.15)}$$

For solution, more equations are required. These are obtained from materials balance, e.g.

$$n_T = n_{CO} + n_{CO_2} + n_{H_2} + n_{H_2O} + n_{N_2} \tag{E.8.16}$$

Let us consider 1 mole of the gas mixture at room temperature. Therefore, the number of moles at room temperature (i.e. initial) are

$$n_{CO}^i = 0.2, \quad n_{CO_2}^i = 0.2, \quad n_{H_2O}^i = 0, \quad n_{H_2}^i = 0.1, \quad n_{N_2}^i = 0.5 \tag{E.8.17}$$

From the *Law of Conservation of Elements*, the initial number of moles of each element is the same as that after equilibrium. Hence,

$$n_C = n_{CO} + n_{CO_2} = 0.4 \quad \text{(from carbon balance)} \tag{E.8.18}$$

$$n_O = n_{CO} + n_{H_2O} + 2n_{CO_2} = 0.6 \quad \text{(from oxygen balance)} \tag{E.8.19}$$

$$n_H = 2n_{H_2} + 2n_{H_2O} = 0.2 \quad \text{(from hydrogen balance)} \tag{E.8.20}$$

$$n_{N_2} = 0.5 \quad \text{(from nitrogen balance)} \tag{E.8.21}$$

From Eqs. (E.8.18)–(E.8.20), through the process of elimination,

$$n_{CO} = 0.4 - n_{CO_2}, \quad n_{H_2O} = 0.2 - n_{CO_2}, \quad n_{H_2} = n_{CO_2} - 0.1 \tag{E.8.22}$$

$$\Delta G^0 \text{ (for E.8.15)} = -RT \ln K_{15} = -RT \ln \left[\frac{p_{CO} \times p_{H_2O}}{p_{CO_2} \times p_{H_2}} \right]_{eq} \tag{E.8.23}$$

Again,
$$\Delta G^0 \text{ (for E.8.15)} = \Delta G^0 \text{ (E.8.1)} - \Delta G^0 \text{ (E.8.5)} \tag{E.8.24}$$
from Hess' Law.

From the values in Eqs. (E.8.2) and (E.8.6) at 900°C,

$$\Delta G^0 \text{ (E.8.15)} = -1.05 \text{ kJ mol}^{-1}, \quad \text{giving } K_{15} = 1.11$$

Hence,

$$\left[\frac{p_{CO} \times p_{H_2O}}{p_{CO_2} \times p_{H_2}} \right]_{eq} = 1.11 \tag{E.8.25}$$

Again, from Eqs. (7.1) and (7.4),

$$p_i = X_i P_T = \frac{n_i}{n_T} \cdot P_T \tag{E.8.26}$$

Combining Eqs. (E.8.22), (E.8.25) and (E.8.26), we obtain

$$\left[\frac{(0.4 - n_{CO_2})(0.2 - n_{CO_2})}{n_{CO_2} \times (n_{CO_2} - 0.1)} \right]_{eq} = 1.11 \tag{E.8.27}$$

Equation (E.8.27) is a quadratic equation. Its solution yields $n_{CO_2} = 0.157$ mole at equilibrium, i.e. the gas contains 15.7% of CO_2 by volume. From Eqs. (E.8.21) and (E.8.22), volume per cent of other constituents at equilibrium at 900°C is obtained, as follows:

CO_2: 15.7%, CO: 24.3%, H_2O: 4.3%, H_2: 5.7%, N_2: 50%   (Ans.)

8.3 Chemical Equilibria Involving Pure Condensed Phases and Gas

From the definition of activity in Eq. (7.15), $a_i = 1$ if the component i is at its standard state. For a solid or liquid, the standard state is pure substance at its stablest state at the temperature where the process takes place. For example, $a_{H_2O} = 1$ for pure ice (below 0°C), and $a_{H_2O} = 1$ for pure water between 0°C and 100°C. Again, for this situation, we do not require any special knowledge of activity vs. composition relationship. In this section, a few examples of equilibria involving such pure condensed phases and gas will be presented as illustrations.

8.3.1 Clausius–Clapeyron Equation—Alternative derivation

The two alternative forms of this equation for vaporization/sublimation of a pure substance A (element or compound) have been derived in Section 6.4.2. Here, these will be derived by employing the equilibrium constant.

Consider the following equilibrium at temperature T:

$$A \text{ (liquid)} = A \text{ (vapour)} \tag{8.3}$$

$$\Delta G_3^0 = -RT \ln K_3 = -RT \ln \left[\frac{p_A}{a_A} \right]_{eq} \tag{8.4}$$

where p_A is vapour pressure of A at temperature T. By convention, $[p_A]_{eq}$ is designated as p_A^0 for pure liquid A. a_A is activity of A in liquid and, by definition, $a_A = 1$ for pure liquid A. Hence,

$$K_3 = p_A^0 \tag{8.5}$$

Variation of equilibrium constant with temperature is given by Van't Hoff equations, viz. Eqs. (7.47) and (7.48). Combining Eq. (8.5) with these, we obtain

$$\frac{d(\ln p_A^0)}{dT} = \frac{\Delta H^0}{RT^2} \tag{8.6}$$

$$\frac{d(\ln p_A^0)}{d(1/T)} = -\frac{\Delta H^0}{R} \tag{8.7}$$

Equations (8.6) and (8.7) are the same as the alternative forms of Clausius–Clapeyron equation, as already derived in section 6.4.2.

8.3.2 Oxidation-reduction Equilibria Involving Pure Metal (M) and Pure Metal Oxide (MO)

Consider the following reaction at temperature T:

$$2M(s, l) + O_2(g) = 2MO(s, l) \tag{8.8}$$

where (s, l) denotes that M and MO may be present as solid or liquid.

By the convention already adopted in this text, ΔG^0 and K for reaction (8.8) would be designated as ΔG_8^0, K_8.

$$\Delta G_8^0 = 2\,\Delta G_f^0\,(MO) = -RT \ln K_8 = -RT \ln \left[\frac{a_{MO}^2}{a_M^2 \cdot p_{O_2}} \right]_{eq} \tag{8.9}$$

where $\Delta G_f^0\,(MO)$ is standard free energy of formation of MO.

If M, MO are pure, then $a_M = 1 = a_{MO}$. Hence, from Eq. (8.9)

$$\Delta G_8^0 = 2\,\Delta G_f^0 = RT \ln \left\lfloor p_{O_2} \right\rfloor_{eq} \tag{8.10}$$

where $\left\lfloor p_{O_2} \right\rfloor_{eq}$ is p_{O_2} in equilibrium with pure M and pure MO.

Phase stability diagram for Cu-Cu₂O-O₂

For the reaction

$$4Cu(s) + O_2(g) = 2Cu_2O(s) \tag{8.11}$$

$$\Delta G_{11}^0 = -3.39 \times 10^5 - 14.25T \ln T + 247T \text{ joules} \tag{8.12}$$

From Eqs. (8.10) and (8.12),

$$\ln [p_{O_2}]_{eq} = -\frac{4.076 \times 10^4}{T} - 1.713 \ln T + 29.69 \tag{8.13}$$

Figure 8.1 presents this relation graphically. On the equilibrium p_{O_2} line, both pure Cu and pure Cu₂O co-exist with O₂. Below the line only Cu is stable solid, and above the line, Cu₂O is stable solid. This can be predicted from common sense. However, for quantitative prediction, Eq. (7.26) is to be employed. For reaction (8.11),

$$K = \left[\frac{1}{p_{O_2}} \right]_{eq}, \quad J = \left[\frac{1}{p_{O_2}} \right]_{actual}$$

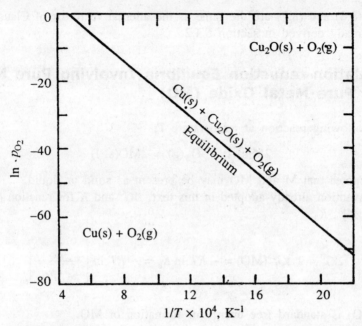

Fig. 8.1 Relative stabilities of Cu and Cu_2O as function of p_{O_2} and temperature.

Substituting these in Eq. (7.27), we get

$$\frac{J}{K} = \frac{(p_{O_2})_{eq}}{(p_{O_2})_{actual}}$$

(8.14)

If $(p_{O_2})_{actual} < (p_{O_2})_{eq}$, $J/K > 1$, and oxidation of Cu is not possible. So, Cu is stable solid below the equilibrium line. The reverse is true above the line.

EXAMPLE 8.4 Determine the lowest temperature at which copper oxide (Cu_2O) can dissociate in a vacuum of 10^{-5} mm Hg.

Solution For dissociation of Cu_2O, ΔG for reaction (8.11) should be positive.

Noting that: $\Delta G = \Delta G^0 + RT \ln J$

and J for reaction (8.11) is

$$\left(\frac{1}{p_{O_2}}\right)_{actual} = \frac{1}{10^{-5}/760}$$

and combining this with Eq. (8.12), the thermodynamic criterion for dissociation is

$$-3.39 \times 10^5 - 14.25T \ln T + 247T + 8.314T \ln (760 \times 10^5) > 0$$

i.e.

$$-3.39 \times 10^5 - 14.25T \ln T + 397.87T > 0 \qquad \text{(E.8.28)}$$

In Eq. (E.8.28), the coefficient of T is $(397.87 - 14.25 \ln T)$. Since $\ln T$ is of the order of 7 to 8 maximum, the coefficient of T is positive (actually, it has to be positive for a valid solution). Hence, the higher the value of T, the more positive will be the LHS. Hence, there will be a lowest temperature $(T_{\min})$ above which dissociation of Cu_2O into Cu and O_2 occurs. At $T = T_{\min}$,

$$\text{LHS} = -3.39 \times 10^5 - 14.25T_{\min} \ln T_{\min} + 397.87T_{\min} = 0 \qquad \text{(E.8.29)}$$

This equation is implicit and can be solved graphically, or numerically through an iterative procedure, or simply by trial and error. The solution of Eq. (E.8.29) yields a value of $T_{\min} = 1140$ K. Therefore, Cu_2O will dissociate at $T > 1140$ K.

EXAMPLE 8.5 The thin layer of silicon can be deposited over a metallic substrate through vapour deposition technique at high temperature, e.g.

$$SiCl_4\,(g) + 2H_2\,(g) = Si\,(s) + 4HCl\,(g) \qquad \text{(8.15)}$$

Assume the temperature is 1200°C and pressure 1 atm. Calculate partial pressures of $SiCl_4$, H_2 and HCl gas at equilibrium with pure solid silicon. Assume that the molar ratio of $SiCl_4 : H_2$ in inlet gas is 1:2, as per stoichiometry of the reactants. ΔG^0 of the reaction is -4183 J mol^{-1} at this temperature (Johnson and Stracher, Chapter 3).

Solution

$$K = \frac{p_{HCl}^4}{p_{H_2}^2 \cdot p_{SiCl_4}} = \exp\left(-\frac{\Delta G^0}{RT}\right) = \exp\left(\frac{4183}{8.314(1200 + 273)}\right) = 1.4 \qquad \text{(E.8.30)}$$

Assume the initial number of moles introduced as 1 mole $SiCl_4$ gas and 2 moles of H_2 gas. Let, upon attainment of equilibrium, n moles of $SiCl_4$ the reacted. Then the situation would be as follows:

Gas	$\rightarrow$	$SiCl_4$	H_2	HCl
Initial moles		1	2	0
Moles reacting or produced		n	$2n$	$4n$
Moles remaining		$1 - n$	$2 - 2n$	$4n$

Since the total number of moles remaining $= (1 - n) + (2 - 2n) + 4n = 3 + n$, and $P_T = 1$ atm, we have, at equilibrium,

$$p_{SiCl_4} = \frac{1 - n}{3 + n}, \qquad p_{H_2} = \frac{2 - 2n}{3 + n}, \qquad p_{HCl} = \frac{4n}{3 + n} \qquad \text{(E.8.31)}$$

Combining Eqs. (E.8.30) and (E.8.31), we get

$$\frac{\left(\dfrac{4n}{3+n}\right)^4}{\left[\dfrac{2-2n}{3+n}\right]^2 \dfrac{1-n}{3+n}} = 1.4 \qquad (E.8.32)$$

i.e.

$$65.4n^4 - 8.4n^2 + 11.2n - 4.2 = 0 \qquad (E.8.33)$$

Solving graphically or by iteration, $n = 0.368$.

Putting in the value of n into Eq. (E.8.31), we obtain

$$p_{SiCl_4} = 0.188 \text{ atm}, \qquad p_{H_2} = 0.375 \text{ atm}, \qquad p_{HCl} = 0.437 \text{ atm}, \qquad P_T = 1 \text{ atm (check)}$$

Thermodynamic basis of finding p_{O_2}, p_{CO}/p_{CO_2} ratios in equilibrium with metal + metal oxide from Ellingham diagram

In the oxide Ellingham diagram (Fig. 7.1), there are three scales around the basic diagram. These are marked as p_{O_2}, p_{H_2}/p_{H_2O}, p_{CO}/p_{CO_2}. These are logarithmic scales and are employed to quickly find out the above gas composition parameters in equilibrium with a metal and its oxide.

The procedure is illustrated with the help of Fig. 8.2 for p_{O_2} and p_{CO}/p_{CO_2}. Let us find out the gas compositions in equilibrium with Mn (s) + MnO (s) at 1200°C, as an example.

There are two points, marked O, C on 0 K vertical line in the Ellingham diagram. The point M is on Mn-MnO line at the desired temperature. If the straight line joining O and M is extrapolated, it intersects at point A on p_{O_2} scale. Point A gives the value of p_{O_2} (in atm) at equilibrium with Mn + MnO at 1200°C. It is approximately 5×10^{-20} atm. Similarly, extrapolation of CM yields point B, which gives the value of p_{CO}/p_{CO_2} ratio in equilibrium with Mn + MnO at 1200°C. Its value is approximately 3×10^4. Similarly, for p_{H_2}/p_{H_2O} ratio, point H in Fig. 7.1 is to be used with a similar procedure. The thermodynamic basis for finding p_{O_2} is Eq. (8.10), from which it follows that

$$\log (p_{O_2})_{eq} \propto \frac{\Delta G_f^0}{T} \propto \frac{PM}{OP} \propto \frac{AQ}{OQ} \qquad (8.16)$$

Since OQ is fixed, $AQ \propto \log(p_{O_2})_{eq}$, and with a calibrated scale, AQ gives the actual value of p_{O_2} in equilibrium with Mn + MnO.

Next, consider the reaction

$$2Mn (s) + 2CO_2 (g) = 2MnO (s) + 2CO (g) \qquad (8.17)$$

Equation (8.17) is obtained by combining the following equations:

$$2Mn (s) + O_2 (g) = 2MnO (s) \qquad (8.18)$$

$$2CO (g) + O_2 (g) = 2CO_2 (g) \qquad (8.19)$$

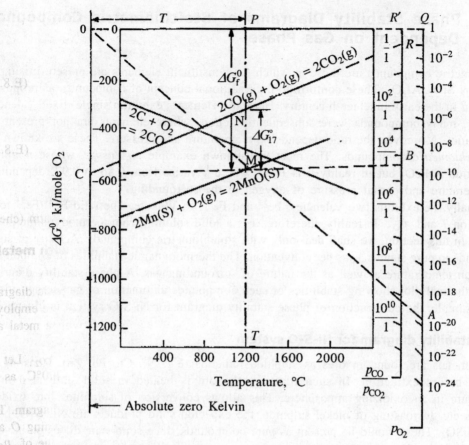

Fig. 8.2 Procedure for finding out p_{O_2} and p_{CO}/p_{CO_2} in equilibrium with a metal and its oxide, from the Ellingham diagram.

From Hess' Law,

$$\Delta G^0_{17} = \Delta G^0_{18} - \Delta G^0_{19} \tag{8.20}$$

ΔG^0_{17} is MN in Fig. 7.2. Again,

$$\Delta G^0_{17} = -RT \ln \left(\frac{p_{CO}}{p_{CO_2}}\right)^2_{eq} = -2RT \ln \left(\frac{p_{CO}}{p_{CO_2}}\right)_{eq} \tag{8.21}$$

Noting that point C lies on CO–CO$_2$ line,

$$\log \left(\frac{p_{CO}}{p_{CO_2}}\right)_{eq} \propto \frac{\Delta G^0_{17}}{T} \propto \frac{MN}{OP} \propto \frac{BR}{OR'} \tag{8.22}$$

Since OR' is fixed, with proper calibration, BR becomes equal to $\log [p_{CO}/p_{CO_2}]_{eq}$.

8.3.3 Phase Stability Diagrams of Stoichiometric Compounds Dependent on Gas Phase

Stoichiometric compounds are those in which the constituent elements are present in simple ratios, e.g. SiO_2, Al_2O_3. These conform to the traditional concept of compounds, as proposed by Dalton in the early nineteenth century. Here, the elements exhibit a single, fixed valency. However, many compounds were subsequently found, where elements are not present in simple ratios. Moreover, the ratio depends on temperature and pressure. These are known as *non-stoichiometric* compounds. The most well-known example is *Wustite*, whose nominal composition is FeO, but in reality it is Fe_xO, where x varies from 0.86 to 0.97, depending on temperature and partial pressure of oxygen in the surrounding.

Actually, Fe exhibits two valencies, Fe^{2+} and Fe^{3+}, in Wustite. The ratio Fe^{2+}/Fe^{3+} ions depends on T and p_{O_2}. In reality, therefore, it is a solid solution from thermodynamic point of view. In this section, we shall deal only with stoichiometric compounds. Activity of such compounds in pure state is 1, as per convention. The thermodynamic stabilities of compounds depend on temperature as well as the nature of surrounding gas. A phase stability diagram is an isothermal plot showing stabilities of such compounds as function of gas composition. As an example, the construction of phase stability diagram for Ni-S-O system is illustrated.

Phase stability diagram for Ni-S-O system

Several metals are found in ores as sulphide minerals (e.g. Ni, Cu, Pb, Zn). Roasting is required before extraction. In such roasting, the ore is heated in solid state at a high temperature in an oxidizing atmosphere. This allows conversion of sulphides into oxides, sulphates etc. In roasting of nickel sulphide, the expected solid products are Ni, NiO, NiS, Ni_3S_2, $NiSO_4$. These would be present as pure compounds since solid-state diffusion is very slow, and solid solution formation does not occur in the finite processing time. The gas is predominantly a mixture of SO_2 and O_2 besides N_2 and which is inert.

The presence or absence of N_2 in gas would not affect the diagram at all. Hence, it is being ignored. The four independent reactions may be written as

$$Ni(s) + \frac{1}{2}O_2(g) = NiO(s) \tag{8.23}$$

$$NiS(s) + \frac{3}{2}O_2(g) = NiO(s) + SO_2(g) \tag{8.24}$$

$$NiO(s) + SO_2(g) + \frac{1}{2}O_2(g) = NiSO_4(s) \tag{8.25}$$

$$3NiS(s) + O_2(g) = Ni_3S_2(s) + SO_2(g) \tag{8.26}$$

The reason these are independent is because each one contains one solid, which is not there in other reactions. Hence, none of the above relations can be obtained by combining the other three reactions.

Since solids are pure, their activities are 1. Hence, the equilibrium relations for the above reactions can be written as

$$\text{Ni–NiO:} \quad \log \ p_{O_2} = -2 \log K_{23} \tag{8.27}$$

$$\text{NiS–NiO:} \quad \log \ p_{SO_2} = \log K_{24} + \frac{3}{2} \log p_{O_2} \tag{8.28}$$

$$\text{NiO–NiSO}_4\text{:} \quad \log \ p_{SO_2} = -\log K_{25} - \frac{1}{2} \log p_{O_2} \tag{8.29}$$

$$\text{NiS–Ni}_3\text{S}_2\text{:} \quad \log \ p_{SO_2} = \log K_{26} + \log p_{O_2} \tag{8.30}$$

Here, p_{SO_2}, p_{O_2} are values in equilibrium with respective solids as shown above. Figure 8.3 presents the phase stability diagram for Ni-S-O system at 1000 K. The plotting is based on Eqs. (8.27)–(8.30). The straight lines correspond to two-phase equilibria. The areas correspond to zones of stability of single solids. Such diagrams were termed as *predominance area diagram* by T.R. Ingraham (1967) who first constructed some of these, including the Ni-S-O system.

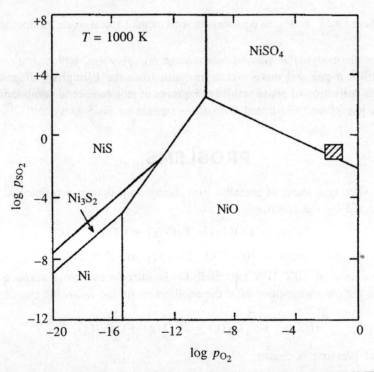

Fig. 8.3 Phase stability diagram for nickel-sulphur-oxygen system at 1000 K [T.R. Ingraham, in *Application of Fundamental Thermodynamics to Metallurgical Processes*, Gordon & Breach, New York, 1967, p. 187].

8.4 Summary

1. For an ideal gas mixture, $X_i = p_i/P_T$, where p_i is the partial pressure of component i, and P_T is the total pressure as given by $P_T = \sum_i p_i$. Again, as derived in Chapter 7, $a_i = p_i$, in an ideal gas mixture.

2. On the basis of the above, and equilibrium relations derived in Chapter 7, some numerical examples of equilibrium calculations involving reactions of ideal gases have been presented. It is to be noted that, in reality too, gaseous equilibria are attained rapidly at high temperatures.

3. From the definition of activity in Chapter 7, $a_i = 1$ if component i is at its standard state, which is pure i at its stablest state at the process temperature, for solids and liquids (i.e. condensed phases). On this basis, some examples of equilibrium relations have been derived and discussed. These are:

    (i)   The derivation of Clausius–Clapeyron equation

    (ii) Pure M + pure metal oxide + O_2 equilibria, for which

$$\Delta G_f^0 \text{ per mole } O_2 = RT \ln \left\lfloor p_{O_2} \right\rfloor_{eq}$$

   where $\left\lfloor p_{O_2} \right\rfloor_{eq}$ is p_{O_2} in equilibrium with metal + metal oxide mixture at temperature T.

    (iii) The thermodynamic basis of determining p_{O_2}, p_{CO}/p_{CO_2} and p_{H_2}/p_{H_2O} in eqilibrium with a metal and metal oxide mixture, from the Ellingham diagram for oxide

    (iv) The derivation of phase stability diagrams of stoichiometric compounds dependent on gas phase, illustrated with the example of Ni-S-O system.

PROBLEMS

8.1 Predict whether a sheet of metallic iron, being annealed in a furnace at 760°C, will get oxidized by the reaction:

$$Fe\ (s) + CO_2\ (g) = FeO\ (s) + CO\ (g)$$

if the atmosphere contains 10% CO, 2% CO_2 and 88% N_2.

8.2 A gas mixture of 50% H_2S and 50% O_2 is introduced into a furnace at 1000 K. Calculate the gas composition after the equilibrium for the following reaction is attained:

$$H_2S\ (g) + \frac{1}{2}O_2\ (g) = H_2O\ (g) + \frac{1}{2}S_2\ (g)$$

The total pressure is 3 atm.

8.3 (a) A gas mixture containing 50% CO_2 and 50% H_2 is introduced into a furnace at 1600°C. What is the value of partial pressure of oxygen in the gas after attainment of equilibrium?

   (b) Using the Ellingham diagram, predict whether this gas will oxidize a piece of titanium kept in the furnace. Note down the procedure you employed.

8.4 How much heat is evolved when 1 mole of SO_2 and 0.5 mole of O_2, each at 1 atm pressure, react to form the equilibrium SO_3–SO_2–O_2 mixture at 1000 K and 1 atm pressure?

8.5 A partial pressure of Cl_2 of 0.1 atm is to be obtained at 500 K and total pressure of 1 atm by establishing equilibrium of the reaction:

$$PCl_5 (g) = PCl_3 (g) + Cl_2 (g)$$

What ratio of $\dfrac{PCl_5}{PCl_3}$ should the initial gas mixture contain?

Given:

$$PCl_3(g) + Cl_2(g) = PCl_5(g)$$

$$\Delta G^0 = -95,600 - 7.94T \ln T + 235.2T \text{ J mol}^{-1}$$

8.6 Calculate the pressure of inert gas which must be applied to liquid lead at 1000°C in order to triple the equilibrium vapour pressure of Pb. The density of liquid Pb at 1000°C is 9.79 g-cm^{-3}.

8.7 Solid MoO_2 is to be reduced by dry hydrogen gas at 1400°C to metallic molybdenum. The total pressure is 1 atm.
   (a) What percentage of H_2 will be utilized for reduction assuming attainment of equilibrium of gas with Mo and MoO_2 mixture?
   (b) Would there be any change if $P_T = 10$ atm?

8.8 Predict whether a gas mixture consisting of 10% H_2S and 90% H_2O can convert solid Cu_2O into solid Cu_2S at 1200 K.

8.9 Predict whether a gas mixture containing 50% H_2, 7% HCl, and 43% Ar at 900 K and 1 atm total pressure is at equilibrium with a mixture of liquid Sn and liquid $SnCl_2$.

8.10 One kilogram of solid calcium carbonate is kept in an evacuated chamber of one cubic meter volume at room temperature, and the system is heated. Calculate:
   (a) The highest temperature up to which solid $CaCO_3$ would be present
   (b) The pressure inside the chamber at 1000 K
   (c) The pressure inside the chamber at 1500 K

8.11 Two of the oxides of iron are FeO and Fe_3O_4. Solid iron exists in equilibrium with one of these oxides at low temperatures and with the other one at high temperatures. Predict which oxide would be at equilibrium with Fe at room temperature and the maximum temperature for this equilibrium co-existence.

8.12 Liquid Mn is kept in a furnace at 1900 K. The furnace atmosphere consists of 2.5% H_2S, 50% H_2, and the remaining N_2. Predict whether MnS will form.

8.13 Solid ZnO and solid ZnS are equilibrated at 2000 K with a H_2S–H_2O–H_2 gas mixture having $p_{H_2O} = 0.5$ and $p_{H_2} = 0.0421$ atm. Calculate the equilibrium partial pressures of O_2, H_2S, S_2 and Zn gas in the atmosphere.

[*Note:* Gas composition is given in volume per cent as per standard convention.]

Chapter 9

Thermodynamics of Solutions

9.1 Introduction

For thermodynamic analysis, we can broadly classify solutions into *aqueous* and *nonaqueous solutions*. Aqueous solutions are extensively dealt with in inorganic and physical chemistry texts. Also, as stated in Chapter 1, metallurgical and materials processing of interest to us are mostly carried out at high temperatures. Here, the solutions are all nonaqueous and inorganic, and hence, these are of special interest to us. Therefore, this chapter covers inorganic nonaqueous solutions, especially for high temperature systems. These may be further classified into:

1. Metallic solutions
2. Non-metallic solutions—oxide solutions, sulphide solutions, halide solutions, etc.

Again, all these may be further sub-classified into *solid solutions* and *liquid solutions*. Single-phase alloys (Ag-Cu, Fe-Ni etc.) are metallic solutions. A molten slag is an oxide solution, and may contain compounds like SiO_2, CaO, and Al_2O_3.

No material (either element or compound) is absolutely pure. It will have some impurities, may be in trace quantities. Even at such low concentrations they may affect some properties significantly. Some examples are now given below:

1. Liquid steel dissolves hydrogen to the extent of few parts per million (ppm). Even this low concentration of H tends to cause fine cracks on surfaces of solid steel during hot forging, thus damaging its properties.
2. Liquid copper dissolves some oxygen during processing. Even a concentration less than 0.1% decreases ductility and electrical conductivity of Cu. Hence, for high conductivity copper, oxygen should be removed to a very low level.
3. In semiconductors, even traces of impurities of the order of parts per billion may affect performance.

9.2 Ideal and Nonideal Solutions

Consider vaporization of component A from a liquid solution at temperature T. If the vapour behaves as an ideal gas, then from Eqs. (8.4) and (8.5), the equilibrium constant is given as

$$K = \frac{p_A}{a_A} = p_A^0 \qquad (9.1)$$

i.e.

$$a_A = \frac{p_A}{p_A^0} \qquad (9.2)$$

where a_A is the activity of component A in the liquid solution. p_A and p_A^0 are equilibrium vapour pressures over the liquid solution and pure A respectively at T.

Equation (9.2) can be understood by imagining two closed containers, kept at temperature T. One contains a liquid solution having species A as a component. The other has pure A. If the containers are kept at T for a long time, vapour-liquid equilibrium will be attained in each of them. In one, vapour pressure of A will be p_A, and in the other, it will be p_A^0. If, for a solution,

$$\frac{p_A}{p_A^0} = X_A, \text{ i.e. } a_A = X_A \qquad (9.3)$$

where X_A is mole fraction of A in the solution, *then the solution obeys Raoult's Law, and is called an ideal solution. Solutions which do not obey Raoult's Law are known as nonideal solutions.* To handle thermodynamics of such solutions, a parameter known as *activity coefficient (γ)* is employed.

$$\gamma_i = \frac{a_i}{X_i}, \text{ i.e. } a_i = \gamma_i X_i \qquad (9.4)$$

where
γ_i = Activity coefficient of component i in a solution, and X_i is mole fraction of i, defined in Eq. (7.1).
If,

$\gamma_i = 1$, solution is ideal.
$\gamma_i > 1$, solution exhibits *positive departure* from Raoult's Law.
$\gamma_i < 1$, solution exhibits *negative departure* from Raoult's Law.

In a solution, $a_i < 1$, in contrast to pure i, where $a_i = 1$. This difference is due to the following two effects:

1. Dilution of component i in a solution due to the presence of other components *(dilution effect)*.
2. Interaction of component i with other components in a solution *(interaction effect)*.

Roughly speaking, activity is a measure of free concentration, i.e. concentration available for reaction. In an ideal solution, the entire concentration is available, i.e. there is only dilution effect.

Interaction effect can be followed more easily if we consider a binary solution A-B. We may visualize three types of bonds between atoms (or molecules) of A and B, viz. A-A, B-B and A-B. Qualitatively speaking, if A-B bonds are stronger than average of A-A and B-B bonds, i.e. A and B have *tendency to form a compound*, then less of A and B atoms would be free. In other words, the entire concentrations of A and B would not be available for reaction. In this case, the solution would exhibit negative departure from Raoult's Law, i.e. $\gamma_A < 1$, $\gamma_B < 1$. In the reverse case, there is tendency *towards formation of clusters* of A and B, and $\gamma_A > 1$ and $\gamma_B > 1$.

Cu-Ag, Fe-Ni, Fe-Mn, FeO-MnO solutions are approximately ideal. Silicates, such as $CaO-SiO_2$, and $MnO-SiO_2$ show negative departures due to their tendency for compound formation. Cu-Fe shows positive departure. These are just some examples. As a generalization, it may be stated that most real solid and liquid solutions are nonideal.

9.3 Partial and Integral Molar Properties; Gibbs–Duhem equation

9.3.1 Relations amongst Partial Molar Properties

Partial molar properties have already been defined and explained in section 7.2.2. Partial molar quantities of mixing also have been defined there. It may be noted here further that all thermodynamic relations derived amongst extensive properties in Chapters 2–5, are applicable here also. For example, in analogy with Eqs. (5.12) and (5.14), we may write

$$d\bar{H}_i = T\,d\bar{S}_i + \bar{V}_i\,dP \qquad (9.5)$$

$$d\bar{G}_i = -\bar{S}_i\,dT + \bar{V}_i\,dP \qquad (9.6)$$

$$d\bar{H}_i^{\mathrm{m}} = T\,d\bar{S}_i^{\mathrm{m}} + \bar{V}_i^{\mathrm{m}}\,dP \qquad (9.7)$$

$$d\bar{G}_i^{\mathrm{m}} = -\bar{S}_i^{\mathrm{m}}\,dT + \bar{V}_i^{\mathrm{m}}\,dP \qquad (9.8)$$

Similarly,

$$\bar{G}_i = \bar{H}_i - T\bar{S}_i \qquad (9.9)$$

$$\bar{G}_i^{\mathrm{m}} = \bar{H}_i^{\mathrm{m}} - T\bar{S}_i^{\mathrm{m}} \qquad (9.10)$$

From section 7.2.2,

$$\bar{G}_i = \left(\frac{\delta G'}{\delta n_i}\right)_{T,P,n_1,n_2,\ldots,\text{except }n_i} \quad , \quad \bar{S}_i = \left(\frac{\delta S'}{\delta n_i}\right)_{T,P,n_1,n_2,\ldots,\text{except }n_i} \qquad (9.11)$$

$$\bar{H}_i = \left(\frac{\delta H'}{\delta n_i} \right)_{T,P,n_1,n_2,\,...,\,\text{except}\, n_i} \quad , \quad \bar{V}_i = \left(\frac{\delta V'}{\delta n_i} \right)_{T,P,n_1,n_2,\,...,\,\text{except}\, n_i} \tag{9.12}$$

$$\bar{G}_i^m = \bar{G}_i - G_i^0, \quad \bar{S}_i^m = \bar{S}_i - S_i^0, \quad \bar{H}_i^m = \bar{H}_i - H_i^0, \quad \bar{V}_i^m = \bar{V}_i - V_i^0 \tag{9.13}$$

where n_1, n_2, ..., n_i denote number of moles of components 1, 2, ..., i, ... etc. in the solution.

9.3.2 Gibbs–Duhem Equation

At constant T and P, as shown in (Eq. (7.7)),

$$dQ' = \left(\frac{\delta Q'}{\delta n_1} \right)_{n_2,n_3,...,n_k\,\text{except}\,n_1} dn_1 + \left(\frac{\delta Q'}{\delta n_2} \right)_{n_1,n_3,...,n_k\,\text{except}\,n_2} dn_2 + ... + \left(\frac{\delta Q'}{\delta n_k} \right)_{n_1,n_3,...\,\text{except}\,n_k} dn_k$$

$$= \bar{Q}_1\, dn_1 + \bar{Q}_2\, dn_2 + ... + \bar{Q}_k\, dn_k$$

$$= \sum_i \bar{Q}_i\, dn_i \tag{9.14}$$

From Eq. (7.1),

$$n_T = n_1 + n_2 + ... + n_i + ... + n_k = \sum_i n_i \tag{9.15}$$

Imagine preparation of this solution. One method of preparation is to keep adding small quantities of components (dn_1, dn_2, ...) in stages at constant T and P such that the overall composition of the solution remains the same in all stages. Since the composition is always constant, the values of $\bar{Q}_1, \bar{Q}_2, ..., \bar{Q}_k$ would also remain constant. This is because these are functions of T, P and composition of solution only.

Therefore, after the additions are complete, we get

$$Q' = \bar{Q}_1 n_1 + \bar{Q}_2 n_2 + ... + \bar{Q}_k n_k = \sum_i \bar{Q}_i n_i \tag{9.16}$$

Since Q is a state property, it does not matter as to how the solution is actually prepared. It is the final state that matters. Hence, Eq. (9.16) will be always valid. Differentiating Eq. (9.16), we get

$$dQ' = (\bar{Q}_1\, dn_1 + n_1\, d\bar{Q}_1) + (\bar{Q}_2\, dn_2 + n_2\, d\bar{Q}_2) + ... + (\bar{Q}_k\, dn_k + n_k\, d\bar{Q}_k) \tag{9.17}$$

Rearranging Eq. (9.17), we obtain

$$dQ' = (\bar{Q}_1\, dn_1 + \bar{Q}_2\, dn_2 + ... + \bar{Q}_k\, dn_k) + (n_1\, d\bar{Q}_1 + n_2\, d\bar{Q}_2 + ... + n_k\, d\bar{Q}_k)$$

$$= \sum_i \bar{Q}_i\, dn_i + \sum_i n_i\, d\bar{Q}_i \tag{9.18}$$

Equating Eqs. (9.14) and (9.18),

$$\sum_i n_i \, d\bar{Q}_i = 0 \tag{9.19}$$

Dividing Eq.(9.19) by n_T, we get

$$\sum_i X_i \, d\bar{Q}_i = 0 \tag{9.20}$$

Equations (9.19 and (9.20) are alternative forms of the Gibbs–Duhem equation (G–H equation), *which provide the principal foundation to thermodynamics of solutions.*

For Gibbs free energy, the G–H equation may be written as

$$\sum_i n_i \, d\bar{G}_i = 0 \quad \text{or} \quad \sum_i X_i \, d\bar{G}_i \tag{9.21}$$

For pure components, the values of Q, viz. $Q_1^0, Q_2^0, ..., Q_k^0$ are constants at constant T and P. So, $dQ_1^0, dQ_2^0, ..., dQ_k^0$ are zero. Hence,

$$\sum_i n_i \, dQ_i^0 = 0, \qquad \sum_i X_i \, dQ_i^0 = 0 \tag{9.22}$$

Subtracting Eq. (9.22) from Eq. (9.19) or (9.20), and from definition of $\bar{Q}_i^m$ in Eq. (7.12), we get

$$\sum_i n_i d(\bar{Q}_i - Q_i^0) = 0, \quad \text{i.e. } \sum_i n_i \, d\bar{Q}_i^m = 0, \qquad \sum_i X_i \, d\bar{Q}_i^m = 0 \tag{9.23}$$

9.3.3 Integral Molar Quantities for Mixing

Dividing Eq. (9.16) by n_T, we obtain

$$Q = X_1\bar{Q}_1 + X_2\bar{Q}_2 + ... + X_k\bar{Q}_k = \sum_i X_i\bar{Q}_i \tag{9.24}$$

where *Q is integral molar value,* as already stated (Eq. 7.5).

Suppose we take X_1 moles of pure component 1, X_2 moles of pure component 2, and so on. Then we prepare a total of 1 mole of mechanical mixture of these pure components. In such a case,

$$Q^0 = X_1Q_1^0 + X_2Q_2^0 + ... + X_kQ_k^0 = \sum_i X_iQ_i^0 \tag{9.25}$$

where the superscript '0' denotes properties at the respective standard states of components, as already noted in Section 7.2.2.

Subtracting Q^0 from Q, we obtain

$$\Delta Q^m = Q - Q^0 = \sum_i X_i(\bar{Q}_i - Q_i^0) = \sum_i X_i\bar{Q}_i^m \tag{9.26}$$

ΔQ^m is known as the *integral molar value for mixing* of the property Q. This represents the change in the value of Q per mole of solution when pure components are mixed to form the solution. Again, ΔQ^m is a function of composition of the solution besides temperature and pressure.

On the basis of Eq. (9.26), we may write

$$\Delta G^m = G - G^0 = \sum_i X_i \bar{G}_i^m \tag{'9.27}$$

where ΔG^m = integral molar Gibbs free energy of mixing of the solution.

EXAMPLE 9.1 If the integral molar enthalpy of mixing (ΔH^m) of a binary solution is LX_1X_2, where L is a constant independent of composition, find the expressions for $\bar{H}_1^m$ and $\bar{H}_2^m$ in terms of mole fractions.

Solution

$$\bar{H}_1^m = \left[\frac{\delta(\Delta H_m')}{\delta n_1} \right]_{n_2, T, P} \tag{E.9.1}$$

where $\Delta H_m' = (n_1 + n_2) \Delta H^m$. Noting that $X_1 = n_1/(n_1 + n_2)$ and $X_2 = n_2/(n_1 + n_2)$,

$$\Delta H_m' = (n_1 + n_2) L \frac{n_1}{n_1 + n_2} \frac{n_2}{n_1 + n_2} = L \frac{n_1 n_2}{n_1 + n_2} \tag{E.9.2}$$

From Eqs. (E. 9.1) and (E. 9.2),

$$\bar{H}_1^m = L \left[\frac{n_2}{n_1 + n_2} + n_1 n_2 (-1) \times \frac{1}{(n_1 + n_2)^2} 1 \right]$$

$$= \frac{L n_2}{n_1 + n_2} \left(1 - \frac{n_1}{n_1 + n_2} \right) = LX_2 (1 - X_1) = LX_2^2$$

since $X_1 + X_2 = 1$. From symmetry, $\bar{H}_2^m = LX_1^2$.

9.4 Some More Properties of Solutions

9.4.1 General Features

From Section 7.2.2,

$$\bar{G}_i^m = RT \ln a_i \tag{7.16}$$

Combining Eqs. (7.16) and (9.27), we get

$$\Delta G^m = RT \sum_i X_i \ln a_i \tag{9.28}$$

Since $a_i = \gamma_i X_i$,

$$\Delta G^{\mathrm{m}} = RT \left(\sum_i X_i \ln X_i + \sum_i X_i \ln \gamma_i \right) \tag{9.29}$$

and,

$$\frac{\bar{G}_i^{\mathrm{m}}}{T} = R(\ln X_i + \ln \gamma_i) \tag{9.30}$$

The application of Gibbs–Helmholtz equation [Eq. (6.28)] to partial molar quantities leads to

$$\left[\frac{\delta(\bar{G}_i^{\mathrm{m}}/T)}{\delta T} \right]_{P,\,\mathrm{com}} = -\frac{\bar{H}_i^{\mathrm{m}}}{T^2} \tag{9.31}$$

In Eq. (9.31), the composition as also pressure is fixed since partial molar properties depend on composition as well. Since X_i is a composition parameter, and is independent of temperature, we have

$$\left[\frac{\delta(\ln X_i)}{\delta T} \right]_{P,\,\mathrm{com}} = 0 \tag{9.32}$$

Combining Eqs. (9.30) to (9.32), we get

$$\left[\frac{\delta(\ln \gamma_i)}{\delta T} \right]_{P,\,\mathrm{com}} = -\frac{\bar{H}_i^{\mathrm{m}}}{RT^2} \tag{9.33}$$

9.4.2 Ideal Solutions

Ideal solutions constitute a special class, and possess some unique characteristic properties. These are derived as follows:

For an ideal solution, $a_i = X_i$. Hence, from Eq. (7.16)

$$\bar{G}_i^{\mathrm{m}} = RT \ln X_i \tag{9.34}$$

From Eqs. (9.27) and (9.34), therefore,

$$\Delta G^{\mathrm{m}} = RT \sum_i X_i \ln X_i \tag{9.35}$$

Again, from Eq. (9.8),

$$\left[\frac{\delta \bar{G}_i^{\mathrm{m}}}{\delta P} \right]_{T,\,\mathrm{comp}} = \bar{V}_i^{\mathrm{m}} \tag{9.36}$$

It is being emphasized again that $\bar{G}_i^m$, $\bar{V}_i^m$ are functions of composition besides T and P. Thus, constant composition restriction is required in Eq. (9.36), besides temperature.

For ideal solution, from Eq. (9.34), it follows that $\bar{G}_i^m$ is not a function of pressure. So, from Eq. (9.36),

$$\bar{V}_i^m = 0 , \text{ and hence,}$$

$$\Delta V^m = \sum_i X_i \bar{V}_i^m = 0 \tag{9.37}$$

Again, from Eq. (9.34),

$$\frac{\bar{G}_i^m}{T} = R \ln X_i \neq f(T) \tag{9.38}$$

Therefore, for an ideal solution, application of Eq. (9.31) leads to

$$\bar{H}_i^m = 0 , \text{ and hence,}$$

$$\Delta H^m = \sum_i X_i \bar{H}_i^m = 0 \tag{9.39}$$

Further, applying the above relation to Eq. (9.10) in an ideal solution, we get

$$\bar{G}_i^m = -T\bar{S}_i^m , \text{ i.e. } \bar{S}_i^m = -R \ln X_i \tag{9.40}$$

and, hence,

$$\Delta S^m = -R \sum_i X_i \ln X_i \tag{9.41}$$

9.5 Binary Solutions

Binary solutions have two components, designated here as A and B. Due to its relative simplicity, and the fact that thermodynamics of ternary and multicomponent solutions are often extrapolated from respective binaries, all introductory thermodynamics texts are primarily concerned with thermodynamic analysis of binary solutions on the basis of general equations already presented.

9.5.1 Activity vs Mole Fraction; Henry's Law

Figure 9.1 presents activity vs mole fraction curves schematically for binary A-B solution at constant temperature. It shows Raoult's Law (RL) lines for components A and B. Positive and negative departures from Raoult's Law are illustrated for a_B. It may be noted that at small values of X_B, the variation of a_B with X_B is linear. This is the basis for Henry's Law (HL) for binary solutions. Henry's Law may be rigorously stated as follows: If $X_B \rightarrow 0$, then $\gamma_B \rightarrow$ a constant (γ_B^0). In other words, in Henry's Law region for B,

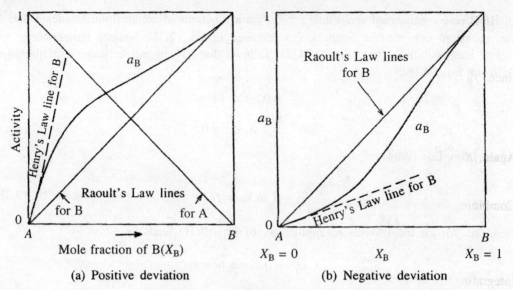

Fig. 9.1   (a) Positive, and (b) negative deviations from the Raoult's Law for component B in the binary A-B solution; Henry's Law lines for B are also shown (schematic).

$$a_B = \gamma_B^0 X_B \qquad (9.42)$$

where γ_B^0 is a constant. B is the solute, and A is known as solvent.

Similarly, if $X_A \to 0$,

$$a_A = \gamma_A^0 X_A \qquad (9.43)$$

One way of classifying solution is division into *dilute* (low concentration of solute) and *concentrated solution*. On this basis, it may also be stated that the *solute in binary dilute solution obeys Henry's Law*. There is no fixed demarcation between dilute and concentrated solutions. From a fundamental point of view, Henry's Law is expected to be obeyed if solute atoms are so far apart from one another that we can ignore solute-solute interactions, i.e. B-B or A-A interactions, as the case may be.

It may also be observed in Fig. 9.1 that the curves for a_B merge with Raoult's Law line if $X_B \to 1$. Similarly curves for a_A merge with Raoult's Law if $X_A \to 1$. In other words, in the region where B obeys Henry's Law, A tends to obey Raoult's Law, and vice versa.

The thermodynamic basis for the above observation is now derived.

From the Gibbs–Duhem equation, viz. Eq. (9.23),

$$X_A d\bar{G}_A^m + X_B d\bar{G}_B^m = 0 \qquad (9.44)$$

i.e.

$$RT[X_A d(\ln a_A) + X_B d(\ln a_B)] = 0 \qquad (9.45)$$

by combining Eq. (9.44) with Eq. (7.16).

If B obeys Henry's Law, then

$$d(\ln a_B) = d[\ln(\gamma_B^0 X_B)] = d[\ln \gamma_B^0 + \ln X_B] = d(\ln X_B) \tag{9.46}$$

since γ_B^0 is a constant. Therefore, from Eqs. (9.45) and (9.46),

$$d(\ln a_A) = -\frac{X_B}{X_A} d(\ln X_B) = -\frac{X_B}{X_A} \frac{dX_B}{X_B} = -\frac{dX_B}{X_A} \tag{9.47}$$

Again, $X_A + X_B = 1$, and so

$$dX_A + dX_B = 0 \tag{9.48}$$

Combining Eqs. (9.47) and (9.48), we get

$$d(\ln a_A) = \frac{dX_A}{X_A} = d(\ln X_A) \tag{9.49}$$

Integrating Eq. (9.49), we obtain

$$\ln a_A = \ln X_A + \text{constant, i.e. } a_A = I X_A \tag{9.50}$$

where I is a constant.

At $X_A = 1$, $a_A = 1$. Hence, $I = 1$, and $a_A = X_A$, i.e. A *obeys Raoult's Law, when B obeys Henry's Law.*

EXAMPLE 9.2 The following partial pressures of Zn have been determined in Cu-Zn alloys at 1060°C:

X_{Zn}	1.0	0.45	0.3	0.2	0.15	0.1	0.05
p_{Zn} (mm Hg)	3040	970	456	180	90	45	22

(i) Does the system obey Raoult's Law?
(ii) In what ranges of composition, does Zn obey Henry's Law?
(iii) What is the free energy change when 1 gm. atom of pure Zn at 1060°C dissolves in a large quantity of the alloy at $X_{Zn} = 0.3$?

Solution

$$a_{Zn} = \frac{p_{Zn}}{p_{Zn}^\circ}, \quad p_{Zn}^0 = 3040 \text{ mm Hg}; \ \gamma_{Zn} = \frac{a_{Zn}}{X_{Zn}}$$

Based on this, the following table is prepared from the data given in the problem:

X_{Zn}	1.0	0.45	0.3	0.2	0.15	0.1	0.05
a_{Zn}	1.0	0.319	0.15	0.0592	0.0296	0.0148	0.0073
γ_{Zn}	1.0	0.709	0.50	0.296	0.197	0.148	0.145

(i) Since γ_{Zn} is not 1 throughout the solution, Raoult's Law is not obeyed.

(ii) In Henry's Law region, γ_{Zn} = constant. It is valid for X_{Zn} less than 0.1 to 0.15 (X_{Zn} < 0.12 as obtained by the graph plotting of γ_{Zn}).

(iii) When 1 g-atom of Zn dissolves in a large quantity of the alloy at 1060°C, the free energy change = $\bar{G}_{Zn} - \bar{G}_{Zn}^0 = \bar{G}_{Zn}^m$ (see Section 7.2.2 and Section 9.3).

$$\bar{G}_{Zn}^m = RT \ln a_{Zn} = RT \ln (0.15) = 5025 \text{ J mol}^{-1} \text{ at } X_{Zn} = 0.3.$$

9.5.2   Variation of ΔG^m, ΔH^m and ΔS^m with Composition for Ideal Binary Solutions

For an ideal solution, $\Delta H^m = 0$ from Eq. (9.39). Further, if the solution is a binary A-B, then from Eqs. (9.35) and (9.41),

$$\Delta G^m = RT(X_A \ln X_A + X_B \ln X_B) \tag{9.51}$$

$$\Delta S^m = - R(X_A \ln X_A + X_B \ln X_B) \tag{9.52}$$

Figure 9.2 shows variations of ΔH^m, ΔS^m and ΔG^m with composition at constant temperature for an ideal binary solution. The curves of ΔG^m, ΔS^m, are symmetric with respect to A and B. ΔS^m is positive and is at maximum at $X_A = X_B = 0.5$. ΔG^m is negative and is at minimum at $X_A = X_B = 0.5$.

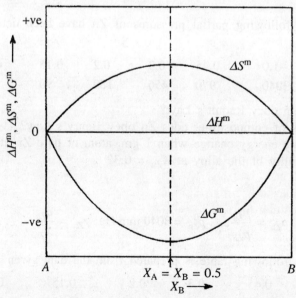

Fig. 9.2   Variations of ΔH^m, ΔS^m, ΔG^m with composition in an ideal binary solution at constant temperature (schematic).

9.5.3 Relationship of $\bar{G}_A^m$, $\bar{G}_B^m$ with ΔG^m in Binary Solutions

For a binary solution, Eq. (9.27) may be rewritten as

$$\Delta G^m = X_A \bar{G}_A^m + X_B \bar{G}_B^m \qquad (9.53)$$

Again, on the basis of Eq. (9.14),

$$d(\Delta G^m) = \bar{G}_A^m \, dX_A + \bar{G}_B^m \, dX_B \qquad (9.54)$$

Combining Eq. (9.54) with Eq. (9.48), we get

$$\frac{d(\Delta G^m)}{dX_A} = \bar{G}_A^m - \bar{G}_B^m \qquad (9.55)$$

Multiplying both sides of Eq. (9.55) by X_B, then adding ΔG^m, and finally combining with Eq. (9.53), we obtain

$$\Delta G^m + X_B \frac{d(\Delta G^m)}{dX_A} = X_B(\bar{G}_A^m - \bar{G}_B^m) + \Delta G^m$$

$$= X_B(\bar{G}_A^m - \bar{G}_B^m) + X_A \bar{G}_A^m + X_B \bar{G}_B^m$$

$$= \bar{G}_A^m (X_A + X_B) = \bar{G}_A^m \qquad (9.56)$$

Similarly,

$$\bar{G}_B^m = \Delta G^m + X_A \frac{d(\Delta G^m)}{dX_B} \qquad (9.57)$$

Figure 9.3 presents a schematic curve of ΔG^m vs. composition for a binary solution at constant temperature. It also demonstrates the graphical procedure to find out $\bar{G}_A^m$ and $\bar{G}_B^m$ at a certain composition from this curve.

By definition, $\Delta G^m = 0$ at $X_A = 1$ and $X_B = 1$ (i.e. for pure A and B). The zero of ΔG^m is at the top of the diagram since ΔG^m for a solution is negative. The point q corresponds to ΔG^m at $X_B = X_B$, i.e. the composition for which $\bar{G}_A^m$ and $\bar{G}_B^m$ are to be determined, sqt is tangent to the curve at q. The graphical procedure gives values of $\bar{G}_A^m$ and $\bar{G}_B^m$ as intercepts of the tangent at $X_A = 1$ and $X_B = 1$, respectively. The basis for the same is explained as follows:

$$os = or + rs = pq + rs$$

$$= \Delta G^m + X_B \frac{d(\Delta G^m)}{dX_A} = \bar{G}_A^m \qquad \text{[from Eq. (9.56)]}$$

$$vt = vy - ty = pq - (-yt) = pq + yt$$

$$= \Delta G^m + X_A \frac{d(\Delta G^m)}{dX_B} = \bar{G}_B^m \qquad \text{[from Eq. (9.57)]}$$

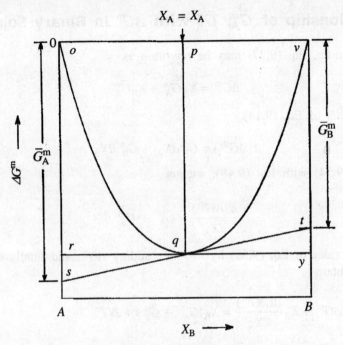

Fig. 9.3 Graphical procedure for determination of $\bar{G}_A^m, \bar{G}_B^m$ from ΔG^m vs. composition curve in a binary solution.

9.6 Gibbs–Duhem Integration for Binary Solutions

Except for the vapour pressure technique with mass spectrometer, experimental arrangement allows measurement of activity of one component only. Let us consider iron-silicon alloy. Let the technique consist of equilibration of the alloy with a gas mixture of $CO + CO_2$ at temperature T. It is found that only SiO_2 is present as the product. It is pure SiO_2 and exists as a separate phase.

On the basis of the above observations, we consider only the following equation:

$$[Si] \text{ in Fe-Si alloy} + 2CO_2(g) = (SiO_2)_{\text{pure solid}} + 2CO(g) \qquad (9.58)$$

$$\Delta G_{58}^0 = -RT \ln K_{58} = -RT \ln \left[\frac{1 \times p_{CO}^2}{[a_{si}]_{\text{alloy}} p_{CO_2}^2} \right]_{eq} \qquad (9.59)$$

since $a_{SiO_2} = 1$.

Rearrangement of Eq. (9.59) yields

$$\ln [a_{si}]_{\text{alloy}} = \frac{\Delta G_{58}^0}{RT} + 2 \ln \left(\frac{p_{CO}}{p_{CO_2}} \right)_{eq} \qquad (9.60)$$

The value of ΔG^0_{58} is available in thermodynamic data books. $\left(\dfrac{p_{CO}}{p_{CO_2}}\right)_{eq}$ is $\dfrac{p_{CO}}{p_{CO_2}}$ ratio in gas at equilibrium with [Si] and SiO_2, and is measured experimentally. Thus, a_{Si} in the alloy is determined with the help of Eq. (9.60).

The question is how do we find out a_{Fe} in the alloy? The answer is that the values of a_{Fe} in the alloy at various concentrations of Si are to be theoretically evaluated from experimental values of a_{Si} as a function of composition of the alloy. The thermodynamic basis for this is Gibbs–Duhem equation.

The G–D equation is a differential equation and is to be integrated. Such integration is straightforward for binary solutions, but involves elaborate procedure for ternary solutions. Since experimental activity versus composition data generally cannot be expressed by equations, the technique involves graph plotting and graphical integration. Nowadays, computer-oriented numerical methods are generally employed.

The method of integration based on Darken's α-function is generally employed. Before discussing this method, the reason for its adoption rather than other simpler methods would be discussed first.

9.6.1 Method 1

Equation (9.44) may be rewritten as the following equation using standard relation between $\bar{G}^m_i$ and a_i:

$$X_A d(\log a_A) + X_B d(\log a_B) = 0 \qquad (9.61)$$

A common logarithm has been chosen in the above equation for convenience of graph plotting.

Integrating Eq. (9.61) for calculating a_A from experimental values of a_B, we get

$$\log a_A \text{ (at } X_A = X_A) = -\int_{\log a_B(\text{at }X_A=1)}^{\log a_B(\text{at }X_A=X_A)} \frac{X_B}{X_A} d(\log a_B) \qquad (9.62)$$

The reason for this lower limit is because at $X_A = 1$, $a_A = 1$ and $\log a_A = 0$. Hence the integration constant becomes zero.

Figure 9.4 is a graph (schematic) of X_B/X_A as function of $\log a_B$. The figure is self-explanatory. At the base line, $X_A = 1$, $X_B = 0$. Hence $a_B = 0$, i.e. $\log a_B \to \infty$. This poses difficulty for proper evaluation of the shaded area. The curve also tends to be asymptotic to the y-axis towards the left-hand side, since $X_A \to 0$, $X_B/X_A \to \infty$. For these reasons, this method is not used.

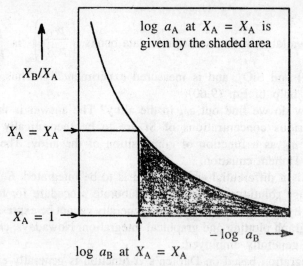

Fig. 9.4 Variation of log a_B with X_B/X_A in a binary solution, and calculation of log a_A by Gibbs–Duhem integration (schematic).

9.6.2 Method 2

Equation (9.48) may be rewritten as

$$\frac{1}{2.303}\left(X_A\frac{dX_A}{X_A} + X_B\frac{dX_B}{X_B}\right) = X_Ad(\log X_A) + X_Bd(\log X_B) = 0 \qquad (9.63)$$

Combining Eq. (9.61) with Eq. (9.63), we obtain

$$X_Ad(\log a_A - \log X_A) + X_Bd(\log a_B - \log X_A) = 0 \qquad (9.64)$$

i.e.

$$X_Ad\log\left(\frac{a_A}{X_A}\right) + X_Bd\left(\log\frac{a_B}{X_B}\right) = X_Ad(\log \gamma_A) + X_Bd(\log \gamma_B) = 0 \qquad (9.65)$$

Integrating Eq. (9.65), we get

$$\log \gamma_A \text{ (at } X_A = X_A) = -\int_{\log \gamma_B(\text{at } X_A = 1)}^{\log \gamma_B(X_A = X_A)} \frac{X_B}{X_A}\, d(\log \gamma_B) \qquad (9.66)$$

X_B/X_A plot against log γ_B is similar to Fig. 9.4 having the following features:

- At base line, $X_A = 1$, $X_B = 0$. In view of Henry's Law, log $\gamma_B = $ log γ_B^0, and hence has a finite value. Therefore, there is no problem at the lower limit.
- However, for $X_A \to 0$, $X_B/X_A \to \infty$, and we shall have asymptotic behaviour as in Fig. 9.4. Thus the shaded area will be somewhat erroneous for small values of X_A.

9.6.3 Method 3 (Darken's Method)

Darken's method eliminates the asymptotic behaviour of curve entirely, and so it allows a reliable method of evaluation of the shaded area. It is based on the α-function, defined as

$$\alpha_i = \frac{\ln \gamma_i}{(1 - X_i)^2} \tag{9.67}$$

For binary solution A-B,

$$\alpha_A = \frac{\ln \gamma_i}{(1 - X_i)^2} = \frac{\ln \gamma_A}{X_B^2}, \qquad \alpha_B = \frac{\ln \gamma_B}{X_A^2} \tag{9.68}$$

If $X_A \to 0$, $X_B \to 1$, $\gamma_B \to 1$, $\ln \gamma_B \to 0$. It can be shown that α_B would have finite value for the above, using the L' Hospital theorem. If $X_B \to 0$, $X_A \to 1$, $\gamma_B \to \gamma_B^0$. Hence, α_B is finite at $X_B \to 0$. Similar would be the behaviour of α_A.

Differentiating Eq. (9.68) by parts, we get

$$d(\ln \gamma_A) = 2\alpha_A X_B dX_B + X_B^2 d\alpha_A \tag{9.69}$$

$$d(\ln \gamma_B) = 2\alpha_B X_A dX_A + X_A^2 d\alpha_B \tag{9.70}$$

Combining Eqs. (9.65) and (9.69), we obtain

$$d(\ln \gamma_A) = - \frac{X_B}{X_A} d(\ln \gamma_B)$$

$$= - \frac{X_B}{X_A} (2\alpha_B X_A dX_A + X_A^2 d\alpha_B)$$

$$= - (2X_B \alpha_B dX_A + X_B X_A d\alpha_B) \tag{9.71}$$

Integrating Eq. (9.71) between $X_A = X_A$ to $X_A = 1$, we get

$$\ln \gamma_A = - \int_{X_A = 1}^{X_A = X_A} 2X_B \alpha_B \, dX_A - \int_{\alpha_B \text{ at } X_A = 1}^{\alpha_B \text{ at } X_A = X_A} X_B X_A \, d\alpha_B \tag{9.72}$$

The formula for integration by parts is

$$\int d(xy) = \int y \, dx + \int x \, dy \tag{9.73}$$

where x and y are two variables. On the basis of Eq. (9.73), we have

$$\int X_B X_A \, d\alpha_B = \int d(X_B X_A \alpha_B) - \int \alpha_B \, d(X_B X_A) \tag{9.74}$$

Substituting Eq. (9.74) into Eq. (9.72), and for integration limits $X_A = 1$ to $X_A = X_A$, we get

$$\ln \gamma_A = -\int 2 X_B \alpha_B \, dX_A - \int d(X_B X_A \alpha_B) + \int \alpha_B \, d(X_B X_A)$$

$$= -\int 2 X_B \alpha_B \, dX_A - X_B X_A \alpha_B + \int \alpha_B X_B \, dX_A + \int \alpha_B X_A \, dX_B$$

$$= -\int 2 X_B \alpha_B \, dX_A - X_B X_A \alpha_B + \int \alpha_B X_B \, dX_A - \int \alpha_B X_A \, dX_A$$

$$= - X_B X_A \alpha_B - \int (2 X_B - X_B + X_A) \alpha_B \, dX_A$$

$$= - X_B X_A \alpha_B - \int_{X_A = 1}^{X_A = X_A} \alpha_B \, dX_A \tag{9.75}$$

Figure 9.5 shows the variation of $\dfrac{\log \gamma_{Cd}}{(1 - X_{Cd})^2}$ $\left(\text{i.e. } \dfrac{\alpha_{Cd}}{2.303}\right)$ with X_{Pb} for Cd-Pb binary at 500°C. The finite values of α_{Cd} at $X_{Pb} = 0$ and $X_{Pb} = 1$ are demonstrated. The integral in Eq. (9.75) is the shaded area, and can be reliably found out in the entire composition range. Thus, this method is generally employed for G–D integration.

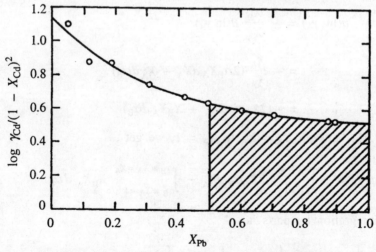

Fig. 9.5 Determination of γ_{Pb} from γ_{Cd} for the Cd-Pb system at 500°C by use of α-Function; the shaded area represents contribution of the final term in Eq. (9.75).

Procedures are also available for G–D integration in ternary solutions, and even quarternaries. However, they are beyond the scope of this text.

EXAMPLE 9.3 In liquid Fe-Ni solution at 1873 K, the activity of nickel as a function of composition is known (noted in the solution that follows).

(i) Calculate the activity of iron at mole fraction of Fe of 0.6 by the Gibbs–Duhem integration.

(ii) Calculate the following at $X_{Fe} = 0.6$: $\bar{G}_{Ni}^m$, $\bar{G}_{Fe}^m$, ΔG^m, G^{XS}.

Solution (i) Given the values of a_{Ni} at various X_{Ni} as follows:

$X_{Ni} =$	0.1	0.2	0.3	0.4	0.5	0.6	0.7	0.8	0.9
$a_{Ni} =$	0.067	0.137	0.208	0.287	0.376	0.492	0.620	0.776	0.89

We have the following calculations:

γ_{Ni}	=	0.67	0.685	0.69	0.72	0.75	0.82	0.986	0.96	0.99
$(1 - X_{Ni})^2 =$		0.81	0.64	0.49	0.36	0.25	0.16	0.09	0.04	0.01
α_{Ni}	=	−0.49	−0.59	−0.75	−0.92	−1.14	−1.24	−1.35	−1.07	−1.01

By method 3 of the Gibbs–Duhem integration (Eq. 9.75),

$$\ln \gamma_{Fe} = - \alpha_{Ni} X_{Fe} X_{Ni} - \int_1^{X_{Fe}} \alpha_{Ni} \, dX_{Fe}$$

The graphical evaluation of the integral (as in Fig. 9.5) yields the value of integral as –0.08. Hence,

$$\ln \gamma_{Fe} = - (- 0.92 \times 0.6 \times 0.4 - 0.08) = 0.301, \text{ i.e.}$$

$$\gamma_{Fe} = 1.35, \qquad a_{Fe} = 0.6 \times 1.35 = 0.81$$

(iii) $\bar{G}_{Ni}^m = RT \ln a_{Ni} = 8.314 \times 1873 \ln (0.237) = -19438 \text{ J mol}$ at $X_{Fe} = 0.6$

$$\bar{G}_{Fe}^m = RT \ln a_{Fe} = 8.314 \times 1873 \ln (0.81) = -3281 \text{ J mol}^{-1}$$

$$\Delta G^m = (X_{Ni} \bar{G}_{Ni}^m + X_{Fe} \bar{G}_{Fe}^m) = 0.4 \times (-19438) + 0.6 (-3281) = -9744 \text{ J mol}^{-1}$$

$$G^{XS} = \Delta G^m - \Delta G^{m, \text{id}} \qquad \text{[as per Eq. (9.80)]}$$

$$= RT(X_{Ni} \ln \gamma_{Ni} + X_{Fe} \ln \gamma_{Fe}) \qquad \text{[see Eq. (9.89]}$$

$$= 8.314 \times 1873 [0.4 \ln (0.72) + 0.6 \ln (1.35)] = 758 \text{ J mol}^{-1}.$$

9.7 Regular Solutions

Ideal solutions provide some equations relating activity to composition, ΔH^m, ΔS^m, ΔG^m expressions. *Most of the solutions, however, are nonideal.* There is a great need of expressing thermodynamic quantities in these solutions in the form of analytical equations as function of composition and temperature.

In this connection, the *regular solution model*, first proposed by Hildebrand, is of great practical utility for various uses. Again, it is an approximation, but is much more consistent with behaviours of a large number of solutions.

Hildebrand defined a regular solution as the one in which

$$\bar{S}_i^m = \bar{S}_i^{m,\,id}, \text{ but } \bar{H}_i^m \neq 0 \tag{9.76}$$

i.e.

$$\Delta S^m = \Delta S^{m,id}, \text{ but } \Delta H^m \neq 0 \tag{9.77}$$

where the superscipt 'id' refers to ideal solution. In other words, a regular solution has the same entropy as an ideal solution of the same composition. But enthalpy is different from ideal solution.

9.7.1 The Excess Function

For dealing with nonideal solutions, a function known as *excess function* (or *molar excess property*) is also employed. It is only for extensive properties, and is defined as

$$Q^{xs} = Q - Q^{id} \tag{9.78}$$

where,

Q = some extensive molar property $(G, H, S, ...)$ of the actual solution
Q^{id} = value of Q for ideal solution at same composition and temperature
Q^{xs} = value of Q in excess of that of the ideal solution.

Similarly, we can define *partial molar excess properties* as

$$\bar{Q}_i^{xs} = \bar{Q}_i - \bar{Q}_i^{id} \tag{9.79}$$

Since the values of Q and $\bar{Q}_i$ are the same (i.e. Q^0, Q_i^0) for standard states for both ideal and nonideal solutions, we may also write

$$Q^{xs} = (Q - Q^0) - (Q^{id} - Q^0) = \Delta Q^m - \Delta Q^{m,id} \tag{9.80}$$

$$\bar{Q}_i^{xs} = (\bar{Q}_i - Q_i^0) - (\bar{Q}_i^{id} - Q_i^0) = \bar{Q}_i^m - \bar{Q}_i^{m,\,id} \tag{9.81}$$

Therefore, on the basis of the above equations, for a regular solution,

$$\bar{S}_i^{xs} = 0, \quad S^{xs} = 0 \tag{9.82}$$

Also,

$$\bar{H}_i^{xs} = \bar{H}_i^m, \quad H^{xs} = \Delta H^m \tag{9.83}$$

since $\bar{H}_i^m$, ΔH^m for ideal solutions are zero.

Hence, for a regular solution,

$$\bar{G}_i^{xs} = \bar{H}_i^{xs} - T\bar{S}_i^{xs} = \bar{H}_i^{xs} = \bar{H}_i^m \tag{9.84}$$

$$G^{xs} = \Delta H^m \tag{9.85}$$

Again,

$$G^{xs} = \Delta G^m - \Delta G^{m, id} = RT \sum_i X_i \ln a_i - RT \sum_i X_i \ln X_i = RT \sum_i X_i \ln \gamma_i \quad (9.86)$$

9.7.2 Regular Binary Solutions—Additional features

Hildebrand showed that α_i, as defined in Eq. (9.67), is not a function of composition in a binary regular solution. If, in Eq. (9.75) α_B is a constant, then

$$\ln \gamma_A = - X_B X_A \alpha_B - \alpha_B(X_A - 1)$$

$$= - X_B X_A \alpha_B + \alpha_B X_B$$

$$= \alpha_B X_B(1 - X_A) = \alpha_B X_B^2 \quad (9.87)$$

Again, from Eq. (9.68), $\ln \gamma_A = \alpha_A X_B^2$. Hence, for a binary regular solution,

$$\alpha_A = \alpha_B = \alpha = \text{a constant} \quad (9.88)$$

Combining Eq. (9.88) with Eq. (9.85) and (9.86), we get

$$\Delta H^m = G^{xs} = RT(X_A \ln \gamma_A + X_B \ln \gamma_B)$$

$$= RT\alpha(X_B^2 X_A + X_A^2 X_B)$$

$$= RT\alpha X_A X_B(X_B + X_A)$$

$$= RT\alpha X_A X_B$$

$$= \Omega X_A X_B \quad (9.89)$$

where Ω is a parameter related to interaction energies in the solution. Again,

$$\bar{H}_A^m = \bar{G}_A^{xs} = RT \ln \gamma_A = RT\alpha X_B^2 = \Omega X_B^2 \quad (9.90)$$

and,

$$\bar{H}_B^m = \bar{G}_B^{xs} = RT \ln \gamma_B = RT\alpha X_A^2 = \Omega X_A^2 \quad (9.91)$$

As discussed earlier, ΔH of reactions and processes mostly do not vary much with temperature. Similar has been the observations on ΔH^m, $\bar{H}_i^m$ of solutions, which have been found to be approximately constant in a limited temperature range (of course, for a fixed composition). Hence, from the above equations, Ω may be assumed to be independent of temperature. Therefore,

$$\alpha = \frac{\Omega}{RT} \propto \frac{1}{T} \quad (9.92)$$

Equation (9.92) is often employed for estimating the effect of temperature on activity etc. where experimental data are not available.

From Eq. (9.89), if ΔH^m *is positive*, then α is positive. Then, on the basis of Eqs. (9.90

and (9.91), $\ln \gamma_A$ and $\ln \gamma_B$ are positive, i.e. $\gamma_A > 1$, $\gamma_B > 1$, i.e. *we have positive departures from Raoult's Law*. Similarly, negative ΔH^m means negative α and negative departures of a_A, a_B from Raoult's Law. Again, *an ideal solution may be looked upon as a regular solution with $\alpha = 0$*.

Figure 9.6 presents variations of ΔH^m, ΔG^m and ΔS^m for some hypothetical regular binary solutions with $\alpha = +1$ and $\alpha = -1$. It also compares these with the ideal solution. Of course, all these curves are at the same temperature.

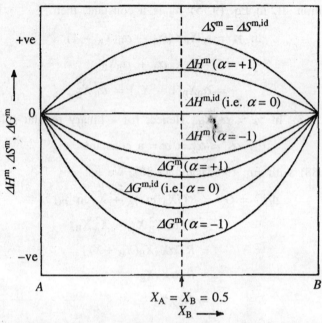

Fig. 9.6 ΔH^m, ΔS^m, ΔG^m as functions of composition at constant temperature for regular binary solutions with different values of α (schematic).

EXAMPLE 9.4 Liquid lead-tin solution exhibits regular solution behaviour. At 746 K, it is given that

$$\ln \gamma_{Pb} = -0.737(1 - X_{Pb})^2$$

(i) What is the corresponding expression for γ_{Sn} at 746 K?

(ii) If 1 mole of solid Pb at 298 K is added to a large quantity of this liquid alloy of composition $X_{Pb} = 0.5$, kept in a thermostat at 746 K, calculate:

    (a) The quantity of heat required to be transferred from the thermostat to bring the final temperature of alloy back to 746 K

    (b) Entropy changes of system and surrounding

    (c) Activity of Pb at 1000 K.

Solution (i) Since it is a regular solution, $\alpha_{Sn} = \alpha_{Pb} = -0.737$, and hence,

$$\ln \gamma_{Sn} = -0.737(1 - X_{Sn})^2$$

We can also derive it by the Gibbs–Duhem integration as follows:

$$\ln \gamma_{Sn} = - X_{Sn}X_{Pb}\alpha_{Pb} - \int_{X_{Sn}=1}^{X_{Sn}} \alpha_{Pb}\,dX_{Sn} \tag{E.9.3}$$

$$= - (-0.737)X_{Sn}X_{Pb} - \int_{1}^{X_{Sn}} (-0.737)\,dX_{Sn}$$

$$= 0.737[X_{Sn}X_{Pb} + (X_{Sn} - 1)]$$

$$= 0.737[X_{Sn}(1 - X_{Sn}) - (1 - X_{Sn})]$$

$$= - 0.737(1 - X_{Sn})^2$$

(ii) (a) Heat to be supplied from the thermostat (ΔH)

= sensible heat required to raise temperature of solid lead + heat of dissolution

$$= \left\lfloor H_{Pb}^0(1)\ (\text{at }746) - H_{Pb}^0(s)\ (\text{at }298) \right\rfloor + \bar{H}_{Pb}^m \tag{E.9.4}$$

$$\bar{H}_{Pb}^m = RT \ln \gamma_{Pb} = 8.314 \times 746 \times (-0.737) \times 0.5^2 = -1142\ \text{J}, \quad \text{at } X_{Pb} = 0.5$$

Since the melting point (T_m) of Pb = 600 K, we have

$$H_{Pb(l)\,746}^0 - H_{Pb(s),\,(298)}^0 = \int_{298}^{600} [C_{P,Pb(s)}]\,dT + \Delta H_m(\text{Pb}) + \int_{600}^{746} [C_{P,Pb(l)}]\,dT \tag{E.9.5}$$

$$= \int_{298}^{600} [23.55 + 9.75 \times 10^{-3}\,T]\,dT + 4212$$

$$+ \int_{600}^{746} [32.43 - 3.1 \times 10^{-3}\,T]\,dT = 17{,}673\ \text{J}$$

Therefore,

$$\Delta H = 17{,}673 - 1142 = 16{,}531\ \text{J}$$

(b) Similarly, ΔS of the alloy as a result of addition is given by

$$\Delta S = \left\lfloor S_{Pb(l),\,746}^0 - S_{Pb(s),\,298}^0 \right\rfloor + \bar{S}_{Pb}^m \tag{E.9.6}$$

$$\bar{S}_{Pb}^m = \bar{S}_{Pb}^{m,\,id} = -R \ln X_{Pb} \quad \text{since it is a regular solution.}$$

Hence,

$$\bar{S}_{Pb}^m = -8.314 \ln (0.5) = 5.76\ \text{J K}^{-1}$$

$$S_{Pb(l),\,746}^0 - S_{Pb(s),\,298}^0 = \int_{298}^{600} \frac{C_{P,Pb(s)}}{T}\,dT + \frac{\Delta H_m(\text{Pb})}{T_m(\text{Pb})} + \int_{600}^{746} \frac{C_{P,Pb(l)}}{T}\,dT \tag{E.9.7}$$

$$= 34.78\ \text{J K}^{-1} \ (\text{after solving})$$

Therefore,

$$(\Delta S)_{alloy} = (\Delta S)_{syst} = 34.78 + 5.76 = 40.54 \text{ J K}^{-1}$$

$$(\Delta S)_{surr} = -\frac{(\Delta H)_{syst}}{746} = -\frac{16,531}{746} = -22.17 \text{ J K}^{-1}$$

i.e. $\quad (\Delta S)_{syst + surr} = 40.54 - 22.17 = +18.37 \text{ J K}^{-1}$

(c) For a regular binary solution, $\alpha \propto 1/T$ (9.92)
Again, at fixed composition, $\ln \gamma_i \propto \alpha$. Hence,

$$\frac{\ln (\gamma_{Pb})_{1000}}{\ln (\gamma_{Pb})_{746}} = \frac{746}{1000}$$

Since at $X_{Pb} = 0.5$, $\gamma_{Pb} = 0.83$, we have γ_{Pb} at 1000 K = 0.88 and $a_{Pb} = 0.44$.

9.8 Quasi-chemical Theory of Solution

Thermodynamic properties of solution are a consequence of interactions amongst atoms/molecules. There have been several attempts to explain the properties from atomistic point of view with varied degrees of success. However, none has been found to be universally acceptable since all of them have some simplifying assumptions. Quasi-chemical theory is the simplest and is able to provide explanations to some behaviour patterns of solutions approximately. Therefore, this is being presented here as an example.

In a metallic solution, the components are assumed to be present as atoms and not molecules due to the nature of metallic bonds. The interaction energy amongst atoms is the sum total of several effects. In quasi-chemical theory, the assumptions are:

1. Atoms have chemical bonds amongst them.
2. The bonds are associated with some enthalpy, which alone contribute to enthalpy of solution.
3. Only bonds amongst nearest neighbouring atoms are energetically significant.

In the binary A-B system, the bonds are A-A, B-B and A-B type. Assume that the enthalpies associated with these bonds are H_{AA}, H_{BB} and H_{AB}. These are enthalpy changes to bring two atoms together from hypothetical gaseous state (i.e. a state where atoms are free). For example, H_{AA} is enthalpy change associated with the following isothermal process:

atom A (g) + atom A (g) = A-A pair in solution

Let us now consider one gram-atom of the solution. Let it have P_{AA}, P_{BB} and P_{AB} numbers of A-A, B-B and A-B bonds, respectively. Then

$$H = P_{AA}H_{AA} + P_{BB}H_{BB} + P_{AB}H_{AB} \qquad (9.93)$$

where H is the enthalpy of formation of the solution for one gram-atom (i.e. per mole) from gaseous state at temperature T.

Let us assume that the solution contains N_A atoms of A and N_B atoms of B. Each A-B bond has one atom of A, whereas each A-A bond has two atoms of A. If Z is the coordination number, then the number of A type atoms in A-B bonds is P_{AB}/Z. On the other hand, the number of A atoms in A-A bonds is $2P_{AA}/Z$. Hence,

$$N_A = \frac{P_{AB}}{Z} + \frac{2P_{AA}}{Z} \qquad (9.94)$$

Similarly,

$$N_B = \frac{P_{AB}}{Z} + \frac{2P_{BB}}{Z} \qquad (9.95)$$

Thus, P_{AA} and P_{BB} can be expressed in terms of P_{AB}, N_A, N_B and Z. Substituting these into Eq. (9.93), we obtain

$$H = \frac{1}{2} ZN_A H_{AA} + \frac{1}{2} ZN_B H_{BB} + P_{AB} \left[H_{AB} - \frac{1}{2}(H_{AA} + H_{BB}) \right] \qquad (9.96)$$

Since absolute thermodynamic quantities cannot be measured, we are interested only in enthalpy of mixing (ΔH^m) of the solution, where

$$\Delta H^m = H - \text{(enthalpy of } N_A \text{ atoms in pure A + enthalpy of } N_B \text{ atoms in pure B)} \qquad (9.97)$$

N_A atoms in pure A contain only A-A bonds, whose numbers are $1/2ZN_A$ since each bond has two A atoms. Similarly, N_B atoms in pure B have $1/2ZN_B$ B-B bonds. Hence,

$$\Delta H^m = H - 1/2ZN_A H_{AA} - 1/2ZN_B H_{BB} \qquad (9.98)$$

Combining Eqs. (9.96) and (9.98), we get

$$\Delta H^m = P_{AB}[H_{AB} - 1/2(H_{AA} + H_{BB})] \qquad (9.99)$$

9.8.1 Application to Ideal and Regular Solutions

Since $\Delta H^m = 0$ for an ideal solution, from Eq. (9.99),

$$H_{AB} = 1/2(H_{AA} + H_{BB}) \qquad (9.100)$$

The physical meaning of this is that the energy of A-B type bond is the arithmetic average of those of A-A and B-B bonds.

The atomistic interpretation of entropy of mixing of ideal solution ($\Delta S^{m,id}$) shall be provided in Chapter 12 on the basis of statistical thermodynamics. This will lead to the conclusion that there is random mixing of atoms in an ideal solution. Since entropy of a regular solution is the same as that of an ideal solution at the same composition and temperature, we may assume random mixing of atoms in a regular solution as well.

If the total number of atoms (i.e. Avogadro's number) in a mole of solution is N_0, then

$$X_A = \frac{N_A}{N_0}, \qquad X_B = \frac{N_B}{N_0} \qquad (9.101)$$

The probability of an A atom to occupy one atomic site is X_A. The probability of a B atom to occupy a specific site is X_B. For random mixing, the joint probability of an A atom and a B atom to occupy neighbouring sites forming A-B bond would, therefore, be $2X_A X_B$. The factor 2 occurs because A and B atoms can interchange sites mutually also. Since the total number of bonds is $1/2 Z N_0$, we have

$$P_{AB} = 2X_A X_B \frac{1}{2} Z N_0 = X_A X_B Z N_0 \qquad (9.102)$$

Substitution of this in Eq. (9.99) yields

$$\Delta H^m = X_A X_B Z N_0 \left[H_{AB} - \frac{1}{2}(H_{AA} + H_{BB}) \right] \qquad (9.103)$$

Comparison of Eq. (9.103) with Eq. (9.89) gives

$$\Omega = Z N_0 \left[H_{AB} - \frac{1}{2}(H_{AA} + H_{BB}) \right] \qquad (9.104)$$

Equation (9.104) thus provides interpretation of Ω parameter of a regular binary solution in terms of atomic parameters.

9.8.2 Non-regular Solutions

In a solid solution, there may be significant strain energy in the lattice due to differences in atomic sizes. In a liquid, strain energy is insignificant and, therefore, quasi-chemical theory is more applicable. For a non-regular solution, Guggenhein and others derived the following equation:

$$P_{AB} = X_A X_B Z N_o \{ 1 - X_A X_B [\exp(2\Omega/(ZRT)) - 1] \} \qquad (9.105)$$

Expanding the exponential and ignoring higher order terms, we get

$$\Delta H^m \cong X_A X_B \Omega [1 - 2X_A X_B \Omega/(ZRT)] \qquad (9.106)$$

$$S^{xs} \approx -X_A^2 X_B^2 \Omega^2/(ZRT^2) \qquad (9.107)$$

$$G^{xs} \approx X_A X_B \Omega [1 - X_A X_B \Omega/(ZRT)] \qquad (9.108)$$

9.9 Summary

1. This chapter deals with thermodynamics of condensed solutions, i.e. solid solutions or liquid solutions. Temperature is assumed constant, unless otherwise stated.

2. A solution, which obeys Raoult's Law, i.e. $a_i = X_i$, is known as an ideal solution.

3. Activity coefficient is defined as: $\gamma_i = \dfrac{a_i}{X_i}$. $\gamma_i = 1$ for ideal solutions, but $\neq 1$ for nonideal solutions.

4. All properties of solutions are functions of composition at a constant temperature.

5. All standard thermodynamic relations derived earlier for pure substances are applicable to properties of solution. For example, in analogy with equation: $dG = -S\,dT + V\,dP$, we may write

$$d\bar{G}_i = -\bar{S}_i\,dT + \bar{V}_i\,dP, \; d\bar{G}_i^m = -\bar{S}_i^m\,dT + \bar{V}_i^m\,dP$$

$$d(\Delta G^m) = -\Delta S^m\,dT + \Delta V^m\,dP$$

6. The alternative forms of the Gibbs–Duhem (G–D) equation are:

$$\sum_i n_i\,d\bar{Q}_i = 0, \quad \sum_i X_i\,d\bar{Q}_i = 0, \quad \sum_i n_i\,d\bar{Q}_i^m = 0, \quad \sum_i X_i\,d\bar{Q}_i^m = 0$$

where Q is an extensive property of state (e.g. H, G, V, S).

7.
$$\Delta Q^m = \sum_i X_i\,\bar{Q}_i^m, \text{ such as } \Delta G^m = \sum_i X_i\,\bar{G}_i^m$$

8. (i) $\dfrac{\delta(\ln \gamma_i)}{\delta T} = -\dfrac{\bar{H}_i^m}{T^2}$ at fixed pressure and composition.

   (ii) $\Delta G^m = RT \sum_i X_i \ln a_i$ at fixed temperature and composition.

9. In an ideal solution, $a_i = X_i$; hence, $\bar{G}_i^m = RT \ln X_i$ and $\Delta G^m = RT \sum_i X_i \ln X_i$. Also,

   $\Delta H^m = 0$, $\Delta V^m = 0$. Therefore, $\Delta S^m = -R \sum_i X_i \ln X_i$.

10. In a binary solution A-B,

    (i) if B is a solute, then at $X_B \to 0$. $\gamma_B \to \gamma_B^0$ (a constant). This is known as Henry's Law;

    (ii) if the solute obeys Henry's Law, then the solvent obeys Raoult's Law;

    (iii) $\bar{G}_A^m = \Delta G^m + X_B \dfrac{d(\Delta G^m)}{dX_A};$

    $\bar{G}_B^m = \Delta G^m + X_A \dfrac{d(\Delta G^m)}{dX_B};$

(iv) Integration of the Gibbs–Duhem equation allows evaluation of a_A(or γ_A) from experimental values of a_B(or γ_B) as a function of composition at constant temperature. Darken's method is generally employed. It uses α-function, where

$$\alpha_i = \frac{\ln \gamma_i}{(1 - X_i)^2}$$

11. In a regular solution, entropy of mixing is equal to that of an ideal solution, i.e.

$$\bar{S}_i^m = \bar{S}_i^{m,\,id}, \qquad \Delta S^m = \Delta S^{m,\,id}$$

However,

$$\bar{H}_i^m \neq 0, \qquad \Delta H^m \neq 0$$

12. Excess property of a solution is defined as

$$Q^{XS} = Q - Q^{id}, \qquad \bar{Q}_i^{XS} = \bar{Q}_i - \bar{Q}_i^{id}$$

Following this, for a regular solution,

$$S^{XS} = 0, \qquad G^{XS} = \Delta H^m = RT \sum_i X_i \ln \gamma_i$$

13. For a regular binary solution A-B,

   (i) $\alpha_A = \alpha_B = \alpha =$ a constant, independent of composition,

   (ii) $G^{XS} = \Delta H^m = RT\alpha X_A X_B = \Omega X_A X_B$.

14. From the quasi-chemical theory of solution,

$$\Omega = ZN_0 \left[H_{AB} - 1/2(H_{AA} + H_{BB})\right]$$

for binary regular solution, where Z is coordination number, N_0 is Avogadro's number, and H_{AB}, H_{AA}, H_{BB} are enthalpies of A-B, A-A, B-B bonds respectively.

PROBLEMS

9.1 Show that in a dilute binary solution A-B at a given temperature, $\bar{H}_B^m$ and $\bar{V}_B^m$ are independent of composition. Here, the subscript B refers to the solute component.

9.2 One mole of solid Cr at 1873 K is added to a large quantity of liquid Fe-Cr solution at 1873 K with $X_{Cr} = 0.2$. If Fe and Cr form ideal solution, calculate the heat and entropy changes in the system resulting from the addition. Ignore the difference of C_P between solid and liquid chromium.

9.3 In a liquid Cu-Sn alloy at 1400 K and at $X_{Sn} = 0.4$, the values of activity of Sn and Cu are 0.333 and 0.362, respectively. Calculate p_{Zn}, p_{Cu}; γ_{Sn}, γ_{Cu}; $\bar{G}_{Sn}^m$, $\bar{G}_{Cu}^m$, ΔG^m, and G^{XS}.

9.4 Evaluate the thermodynamic feasibility of removal of aluminium from liquid silver under vacuum at 1000°C. The maximum residual Al in the solution should be 1 mole percent. Henry's Law constant for $Al(\gamma^0_{Al}) = 0.11$.

9.5 The following data are given for liquid Cu-Sn alloys at 800°C:

$X_{Sn} =$	0.1	0.2	0.3	0.4	0.5	0.6	0.7	0.8	0.9
$a_{Sn} =$	0.007	0.055	0.213	0.353	0.474	0.574	0.683	0.805	0.909
$a_{Cu} =$	0.851	0.602	0.388	0.34	0.267	0.21	0.153	0.098	0.047

Note: a_{Sn} values were experimental; a_{Cu} values calculated by Gibbs–Duhem integration with hypothetical liquid Cu at 800°C as standard state.

(i) Do Sn and Cu obey Raoult's Law?
(ii) Find out the composition ranges for the validity of Henry's Law for Sn and Cu.
(iii) Calculate the following at $X_{Sn} = 0.6$: p_{Sn}, p_{Cu}; $\bar{G}^m_{Sn}$, $\bar{G}^m_{Cu}$, ΔG^m, G^{XS}.
(iv) By the Gibbs–Duhem integration, verify the value of a_{Cu} at $X_{Sn} = 0.6$.
(v) Is it a regular solution?

9.6 The following data are provided for liquid Cu-Ag alloys at 1300 K:

$X_{Cu} =$	0.1	0.2	0.3	0.4	0.5	0.6	0.7	0.8
$\gamma_{Cu} =$	2.54	2.09	1.74	1.50	1.36	1.22	1.14	1.04
$\gamma_{Ag} =$	1.01	1.05	1.11	1.20	1.32	1.50	1.75	2.16

Note: Like Problem 5, γ_{Ag} calculated by GD integration.

The following additional data are also provided at $X_{Cu} = 0.5$:

(i) $\gamma_{Cu} = 1.4$, $\gamma_{Ag} = 1.37$, at 1200 K.
(ii) $\gamma_{Cu} = 1.32$, $\gamma_{Ag} = 1.29$, at 1400 K.

Calculate: (i) γ_{Cu} by GD integration at any of the above values of X_{Cu} (as per your choice) at 1300 K, and verify against data.

(ii) the following at $X_{Cu} = 0.5$ and $T = 1200$ K, 1300 K, 1400 K:

$$\bar{G}^m_{Cu}, \bar{G}^m_{Ag}, \Delta G^m, G^{XS}; \alpha_{Cu}, \alpha_{Ag};$$

(iii) $\bar{H}^m_{Cu}, \bar{H}^m_{Ag}, \bar{H}^{XS}_{Cu}, \bar{H}^{XS}_{Ag}, \Delta H^m, H^{XS}$ at $X_{Cu} = 0.5$;

(iv) $\bar{S}^m_{Cu}, \bar{S}^m_{Ag}, \bar{S}^{XS}_{Cu}, \bar{S}^{XS}_{Ag}, \Delta S^m, S^{XS}$ at $X_{Cu} = 0.5$;

(vi) Compare the characteristics of this solution with various characteristics of a regular solution.

9.7 In liquid Cu-Zn solution,

$$\alpha_{Zn} = -\frac{19,250}{RT}, \text{ where } R \text{ is in J mol}^{-1} \text{ K}^{-1}$$

(i) by GD integration, show that $\alpha_{Cu} = \alpha_{Zn}$;

(ii) calculate G^{XS}, ΔG^m at 1400 K and $X_{Cu} = 0.6$.

9.8 Assume that, in a binary solution A-B, the activity coefficients can be expressed as

$$\ln \gamma_A = \alpha_1 X_B + 1/2\ \alpha_2 X_B^2 + 1/3\ \alpha_3 X_B^3 + ...$$

$$\ln \gamma_B = \beta_1 X_A + 1/2\ \beta_2 X_A^2 + 1/3\ \beta_3 X_A^3 + ...$$

in the entire composition. Prove:

(i) $\alpha_1 = \beta_1 = 0$

(ii) if only the quadratic terms are retained, $\alpha_2 = \beta_2$.

9.9 In the zinc-cadmium solution at 700 K,

$$\ln \gamma_{Cd} = 0.967(1 - X_{Cd})^2 + 0.691\ (1 - X_{Cd})^3$$

(i) Express $\ln \gamma_{Zn}$ as a function of mole fraction.

(ii) Find out Henry's Law constant for zinc.

(iii) If $\ln \gamma_{Cd} = 0.967(1 - X_{Cd})^2$, calculate $\bar{H}_{Cd}^m$ at $X_{Zn} = 0.3$.

Chapter *10*

Chemical Potential and Equilibria amongst Phases of Variable Compositions

10.1 Chemical Potential

10.1.1 Definition

In Section 1.3, we discussed about thermodynamic equilibrium. It was mentioned there that for physico-chemical processes and chemically reactive systems, thermodynamic equilibrium requires attainment of *physico-chemical equilibrium* (also referred to as *chemical equilibrium*) besides mechanical and thermal equilibria. This means that chemical potential should be uniform in the entire system in addition to uniformity of pressure and temperature.

J.W. Gibbs first proposed the term 'chemical potential' in analogy with other potentials (e.g. thermal potential and gravitational potential). The meaning of the term will become clear as we go along.

In Chapter 7, a distinction was made between a molar quantity (Q) and the same for system as a whole (Q') vide Eq. (7.5), for convenience of derivation. By adopting this convention, let us consider a system with variable composition. Then Eqs. (5.11)–(5.14) may be modified as follows:

$$dU' = T\,dS' - P\,dV' + \left(\frac{\delta U'}{\delta n_1}\right)_{S',V',n_2,\,\ldots\,\text{except }n_1} dn_1 + \left(\frac{\delta U'}{\delta n_2}\right)_{S',V',n_1,\,\ldots\,\text{except }n_2} dn_2 + \ldots + \left(\frac{\delta U'}{\delta n_i}\right)_{S',V',n_1,\,\ldots\,\text{except }n_i} dn_i + \ldots$$

$$= T\,dS' - P\,dV' + \sum_i \left(\frac{\delta U'}{\delta n_i}\right)_{S',V',n_1,\,\ldots\,\text{except }n_i} dn_i \tag{10.1}$$

159

Similarly,

$$dH' = T \, dS' + V' \, dP + \sum_i \left(\frac{\delta H'}{\delta n_i} \right)_{S', P, n_1, \, ... \, \text{except} \, n_i} dn_i \qquad (10.2)$$

$$dA' = - S' \, dT - P \, dV' + \sum_i \left(\frac{\delta A'}{\delta n_i} \right)_{T, V', n_1, \, ... \, \text{except} \, n_i} dn_i \qquad (10.3)$$

$$dG' = - S' \, dT + V' \, dP + \sum_i \left(\frac{\delta G'}{\delta n_i} \right)_{T, P, n_1, \, ... \, \text{except} \, n_i} dn_i \qquad (10.4)$$

As defined in section 7.2.1, n_1, n_2, ..., n_i, ... are numbers of moles of components 1, 2, ..., i ... in the solution. The last term in Eqs. (10.1)–(10.4) is related to variation of U', H', A' and G' with change of n_1, n_2 etc. As usual, i is the general symbol of a component.

Gibbs' original definition of chemical potential of a component $i(\mu_i)$ was

$$\mu_i = \left(\frac{\delta U'}{\delta n_i} \right)_{S', V', n_1, n_2 \, ... \, \text{except} \, n_i} \qquad (10.5)$$

Therefore,

$$dU' = T \, dS' - P \, dV' + \sum_i \mu_i \, dn_i \qquad (10.6)$$

Again,

$$dH' = dU' + P \, dV' + V' \, dP \qquad (2.25)$$

Combining Eqs. (10.6) and (2.25), we obtain

$$dH' = T \, dS' + V' \, dP + \sum_i \mu_i \, dn_i \qquad (10.7)$$

Equating relations (10.2) and (10.7), we get

$$\mu_i = \left(\frac{\delta H'}{\delta n_i} \right)_{S', P, n_1, n_2, \, ... \, \text{except} \, n_i} \qquad (10.8)$$

Similarly, by combining Eqs. (10.3) and (10.4) with Eq. (10.6), the following relations can be obtained:

$$\mu_i = \left(\frac{\delta A'}{\delta n_i} \right)_{T, V', n_1, n_2, \, ... \, \text{except} \, n_i} \qquad (10.9)$$

$$\mu_i = \left(\frac{\delta G'}{\delta n_i} \right)_{T, P, n_1, n_2, \, ... \, \text{except} \, n_i} \qquad (10.10)$$

In accordance with the above equations, therefore, there are four different definitions of chemical potential. In chemical thermodynamics, we employ Gibbs free energy criterion for equilibrium. Hence, in this text as well as in other texts dealing with metallurgical thermodynamics, the definition of chemical potential, as given in Eq. (10.10), will be employed. This way, μ_i *becomes the same as the partial molar free energy of component i (i.e. $\bar{G}_i$) in a solution*, as defined and discussed in section 7.2.2.

10.1.2 Equality of Chemical Potentials amongst Phases at Equilibrium at Constant Temperature and Pressure

At constant T and P, Eq. (10.4) may be rewritten as

$$dG' = \mu_1 \, dn_1 + \mu_2 \, dn_2 + \dots + \mu_i \, dn_i + \dots \tag{10.11}$$

Consider 2 phases (I and II) in the system. Then,

$$dG'^{,\mathrm{I}} = \mu_1^{\mathrm{I}} \, dn_1^{\mathrm{I}} + \dots + \mu_i^{\mathrm{I}} \, dn_i^{\mathrm{I}} + \dots \tag{10.12}$$

$$dG'^{,\mathrm{II}} = \mu_1^{\mathrm{II}} \, dn_1^{\mathrm{II}} + \dots + \mu_i^{\mathrm{II}} \, dn_i^{\mathrm{II}} + \dots \tag{10.13}$$

Consider exchange of dn_1 moles of *only component 1* between phases. Then,

$$dG'^{,\mathrm{I}} = \mu_1^{\mathrm{I}} \, dn_1^{\mathrm{I}}, \, dG'^{,\mathrm{II}} = \mu_1^{\mathrm{II}} \, dn_1^{\mathrm{II}} \tag{10.14}$$

Since,

$$dn_1^{\mathrm{II}} = - dn_1^{\mathrm{I}} = - dn_1 \tag{10.15}$$

Therefore,

$$dG' = dG'^{,\mathrm{I}} + dG'^{,\mathrm{II}} = dn_1 (\mu_1^{\mathrm{I}} - \mu_1^{\mathrm{II}}) \tag{10.16}$$

where dG' is the total free energy change in the system as a result of exchange of dn_1 between phases I and II. For equilibrium at constant T and P,
$dG' = 0$, and hence from Eq. (10.16),

$$\mu_1^{\mathrm{I}} = \mu_1^{\mathrm{II}} \tag{10.17}$$

Equation (10.17) can be generalized for all components with similar arguments. Hence, at constant T and P, if phases I and II are at equilibrium, then

$$\mu_1^{\mathrm{I}} = \mu_1^{\mathrm{II}}, \, \mu_2^{\mathrm{I}} = \mu_2^{\mathrm{II}}, \, \dots, \, \mu_i^{\mathrm{I}} = \mu_i^{\mathrm{II}}, \, \dots \tag{10.18}$$

Similarly, Eq. (10.17) can be generalized for any number of phases at equilibrium as

$$\mu_1^{I} = \mu_1^{II} = \mu_1^{III} = \ldots = \mu_1^{P}$$

$$\mu_2^{I} = \mu_2^{II} = \mu_2^{III} = \ldots = \mu_2^{P}$$

$$\vdots \qquad \vdots \qquad \vdots \qquad \qquad \vdots$$

$$\mu_i^{I} = \mu_i^{II} = \mu_i^{III} = \ldots = \mu_i^{P} \tag{10.19}$$

where P is the total number of phases in the system.

Equation (10.19) is a statement of equality of chemical potentials of all components amongst all phases at equilibrium at the same pressure and temperature.

10.1.3 Some Additional Remarks on Chemical Potential

Elucidation of Eq. (10.19)

Equation (10.19) is the most general one. For application purposes, it does not matter whether a phase is a solution or a pure substance. However, it is necessary to clarify some confusions.

Let us consider equilibrium between liquid copper and oxygen gas, and find out the chemical potential of O_2 (μ_{O_2}) in liquid copper. The confusion may be due to the following reason. Oxygen dissolves in metals in the atomic state. How can we imagine μ_{O_2} in liquid copper then?

The answer lies in the fact that, according to equations of chemical equilibria, the concentration of molecular O_2 in the melt may be extremely small, but it can never be zero. This follows from the equilibrium

$$2[O] = [O_2]; \qquad K_{20} = \frac{[a_{O_2}]}{[a_O]^2} \tag{10.20}$$

Here, [ ] denotes components in metallic solution. K_{20} has a nonzero value. If it is zero, then $\Delta G_{20}^0 = -RT \ln K_{20}$ would be infinity, which is not possible. Thus, $[a_{O_2}]$ can not be zero.

To take another example, suppose a piece of pure SiO_2 is brought to equilibrium with a molten Cu-Ag alloy. How can we think of μ_{Ag} in SiO_2 and μ_{Si} in the alloy? The explanation can be given with the help of the following reactions:

$$SiO_2(s) + 4[Ag] = [Si] + 2(Ag_2O) \tag{10.21}$$

$$(Ag_2O) = 2(Ag) + (O) \tag{10.22}$$

$$SiO_2(s) + 4[Cu] = [Si] + 2(Cu_2O) \tag{10.23}$$

In the above equations, ( ) denotes oxide phase. Some Si would dissolve in the alloy, through reactions (10.21) and (10.23). Its concentration would be very small. Similarly, through reactions (10.21) and (10.22), a very very small concentration of Ag in oxide phase can be predicted from thermodynamic equilibria considerations as in the case of O_2 in Eq. (10.20).

μ_i criteria for phases not at equilibrium

If a process is to occur, then $(dG')_{T,P} < 0$. Combining this with Eq. (10.16), the process would take place if

$$\mu_1^I < \mu_1^{II} \qquad (10.24)$$

For Eq. (10.16), the process under consideration is transfer of component 1 from phase II to phase I. [*Note: dn_1 has been taken as positive. Hence dn_1^I is positive, i.e. phase I receives dn_1 and phase II loses it.*]

Equation (10.24) can be generalized to state that a *component would tend to be transferred from a higher to a lower chemical potential. This makes μ_i analogous to thermal potential etc.* For example, heat flows from higher thermal potential (i.e. higher temperature) to lower thermal potential.

Application of chemical potential

Overall chemical equilibrium calculations are carried out through equilibrium constant as illustrated in Chapters 7 and 8. However, there are areas of application where chemical potential approach offers advantages. These are as follows:

1. Studies of phase equilibria will be illustrated in this chapter itself.
2. In multicomponent systems, often we are interested in one component only. Hence, an easy approach is to consider chemical potential of that component alone.
3. Equation (10.24) allows easy visualization as to which way a species (element or compound) would tend to get transferred.

Chemical potential and activity

Equation (7.16) may be rewritten in terms of chemical potential as

$$\mu_i - \mu_i^0 = RT \ln a_i \qquad (10.25)$$

where, $\mu_i^0 = \mu_i$ at standard state of i.

By universal convention, μ_i^0 is taken as zero for all phases in the system. It also means that we adopt the same standard state in all phases. This simplifies Eq. (10.25) into

$$\mu_i = RT \ln a_i \qquad (10.26)$$

For example, for Eq. (10.20), chemical potential of O_2 (also known as *oxygen potential*) in both the gas and liquid copper co-existing at equilibrium may be written as

$$\mu_{O_2} = RT \ln a_{O_2} = RT \ln p_{O_2} \qquad (10.27)$$

where p_{O_2} = partial pressure O_2 in gas at equilibrium.

From Eq. (10.26), we may also conclude that at equilibrium (provided standard states are same in all phases, (say) pure components 1, 2, ..., i, ...),

$$a_1^I = a_1^{II} = ... = a_1^P \qquad (10.28)$$
$$\vdots$$
$$a_i^I = a_i^{II} = ... = a_i^P$$
$$\vdots$$

Also, species i will be transferred from phase I to phase II, if $a_i^I > a_i^{II}$.

10.2 Gibbs Phase Rule and Applications

10.2.1 Derivation of Gibbs Phase Rule

Consider a system at equilibrium with the following specifications:

1. Constant temperature and pressure
2. Components are 1, 2, ..., i, j, ...
3. No. of components = C
4. All components are independent, i.e. they do not enter into chemical reactions with one another
5. No. of phases = P.

The thermodynamic state of each phase would be determined by temperature, pressure and composition variables (mole fraction, weight % or anything else). The number of composition variables to be fixed independently is $(C - 1)$.

Therefore, the thermodynamic state of a single phase would be determined by $(C - 1 + 2)$, i.e. $C + 1$ variables. The number "2" stands for T and P, which constitute two independent variables. For the entire system of P phases, the total number of variables would be $P(C + 1)$.

All the above variables, however, would not be independent since the requirements for thermodynamic equilibrium impose the following *restrictions* (i.e. *constraints*):

$$T^{\mathrm{I}} = T^{\mathrm{II}} = ... = T^{P} \quad \text{for thermal equilibrium}$$

$$P^{\mathrm{I}} = P^{\mathrm{II}} = ... = P^{P} \quad \text{for pressure equilibrium}$$

$$\mu_1^{\mathrm{I}} = \mu_1^{\mathrm{II}} = ... = \mu_1^{P}$$
$$\vdots$$
$$\mu_i^{\mathrm{I}} = \mu_i^{\mathrm{II}} = ... = \mu_i^{P} \quad (C\text{-equations for chemical equilibrium}) \quad (10.29)$$
$$\vdots$$

where I, II, ... are phases. From Eq. (10.29),

$$\text{Total No. of constraints} = (P - 1)(C + 2) \quad (10.30)$$

since there are $(P - 1)$ equalities in each line of Eq. (10.29). Hence,

$$\text{Total No. of independent variables} = F = P(C + 1) - (P - 1)(C + 2)$$
$$= C - P + 2 \quad (10.31)$$

Equation (10.31) is the famous Gibbs phase rule.

10.2.2   Application of Phase Rule

Non-reacting components

EXAMPLE 10.1   *Solid-Liquid-Vapour Equilibria* (Section 6.4). If all three-phases co-exist, then $P = 3$. For a pure substance, $C = 1$. From Eq. (10.31), therefore, $F = 1 - 3 + 2 = 0$. This means an *Invariant equilibrium* with no degree of freedom at all, i.e. we cannot vary T or P independently. This explains the triple point phenomena (see Fig. 6.4).

EXAMPLE 10.2   *Alloy Phase Diagrams.* Here, the system is non-reactive, the environment of the alloy being either vacuum or an inert gas (Ar, He etc.). Again, pressure has hardly any influence. So, Eq. (10.31) gets simplified into

$$F = C - P + 1 \qquad\qquad\qquad (\text{E}.10.1)$$

For a binary system (Cu-Ag, Fe-Ni, etc.), $C = 2$. Hence, $F = 3 - P$. If only liquid or solid solution is present, then $P = 1$ and $F = 2$. This means both temperature and composition can be varied independently. If there is a mixture of solid solution + liquid solution, then $P = 2$ and $F = 1$. This means that if temperature is fixed, compositions of solid and liquid are automatically fixed, and vice versa.

Figure 10.1 shows the binary phase diagram of Pb-Sn system. At 200°C, compositions of liquid as well as α and β solid solutions can be varied over some ranges. But composition

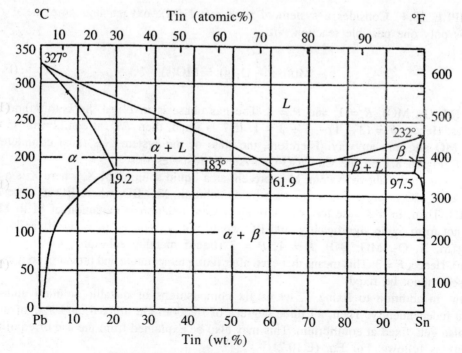

Fig. 10.1   Lead-tin binary phase diagram.

of the liquid solution on liquidus and that of the solid solution on the solidus are fixed since these correspond to equilibrium co-existence of two-phases.

At the *eutectic point,* there are three-phases co-existing at equilibrium. Therefore, $F = 0$ here, and is an invariant.

EXAMPLE 10.3 *Nonreacting Oxide Phase Diagrams.* A good example of non-reacting oxide phase diagram is the SiO_2-Al_2O_3 binary system which constitutes the principal phase diagram for oxide refractory systems (Fig. 10.2). Like alloy systems, here also the gas phase and pressure are neglected (i.e. $F = C - P + 1$).

Again, both temperature and composition of liquid phase can be varied, but at a fixed temperature, liquidus composition is fixed. Solid phases are pure. Point P is an invariant known as *peritectic.* Here, three phases co-exist at equilibrium (liquid, corundum, mullite).

Reacting components

If components are reacting amongst themselves, then the reaction equilibria constitute additional constraints. However, all possible reactions are not independent. Only the number of independent reactions are to be taken into account. In such a situation,

$$C = N - R \tag{10.32}$$

where N is the number of species (elements and compounds) and R is the number of independent reactions.

EXAMPLE 10.4 Consider a system of pure metal (M), oxygen and pure metal oxide. There is only one possible reaction, viz.

$$M(s) + \frac{1}{2}O_2(g) = MO(s) \tag{E.10.2}$$

$N = 3$ (M, O_2, MO), $R = 1$, and $P = 3$. The gas phase is involved. So both T and P are variables. Hence, $F = (3 - 1) - 3 + 2 = 1$. If T is fixed, then p_{O_2} at equilibrium is fixed. M and MO are pure anyway. Therefore, the state of the system gets fixed completely by fixing T. Figure 8.1 on the Cu-Cu_2O-O_2 system illustrates this.

Suppose, in the above system, M is present as a liquid solution M-X, where X is a noble metal. Also, MO is present as a separate oxide solution, MO-BO, where BO is more stable than MO. Then, in this case too, the only significant reaction is oxidation of M to MO. X would not form oxide significantly, and BO would not be reduced to B significantly. Here, $N = 5$ (M, X, O_2, MO, BO), $R = 1$, $P = 3$ (liquid metallic solution, gas, liquid oxide solution). Hence, $F = 3$. This means that even after fixing temperature and pressure, compositions of phases cannot be fixed.

Now, in addition to fixing T, let us fix compositions of metallic solution and oxide solution independently. Then, all three degrees of freedom are utilized, and p_{O_2} of the gas phase also gets fixed at equilibrium. This may also be explained from the use of equilibrium constants as follows. For Eq. (E.10.2),

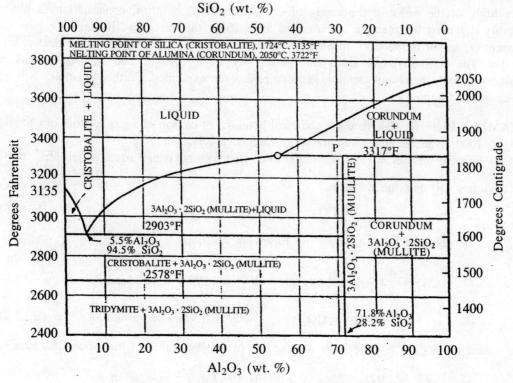

Fig. 10.2 Silica-alumina binary phase diagram.

$$K = \left\{ \frac{(a_{MO})}{[a_M] \left[p_{O_2} \right]^{1/2}} \right\}_{eq} \qquad \text{(E.10.3)}$$

K is fixed by temperature, a_{MO} and a_M are fixed by temperature and composition. Thus, from Eq. (E.10.3), $(p_{O_2})_{eq}$ is automatically fixed.

EXAMPLE 10.5 Dissociation of limestone

$$CaCO_3(s) = CaO(s) + CO_2(g) \qquad \text{(E.10.4)}$$

$N = 3$ ($CaCO_3$, CaO, CO_2), $R = 1$ (i.e. Eq. (E.10.4), $P = 3$ ($CaCO_3$, CaO, gas).

Hence, $F = (3 - 1) - 3 + 2 = 1$. If temperature is fixed, then $(p_{CO_2})_{eq}$ is automatically fixed, thus fixing the state of the entire system.

Figures. 10.1 and 10.2 have presented phase diagrams of Pb-Sn and SiO_2-Al_2O_3 binary systems, respectively. These are examples of condensed phase systems (i.e. solids and liquids), which are nonreactive. The influence of pressure is also negligible (unless the pressure is very high, of the order of thousands of atmospheres). In contrast, construction of phase stability diagrams for reactive systems has to consider the gas phase if there are gaseous reactant (s) and/or product (s). In Section 8.3.2, we have already provided some discussions on this. The thermodynamic basis for finding out p_{O_2}, p_{CO}/p_{CO_2} and p_{H_2}/p_{H_2O} ratios at equilibrium with metal and metal oxide also have been explained. Further examples will be presented in Chapter 11.

EXAMPLE 10. ⟩   (i) Calculate the chemical potential of carbon of a gas mixture of CO and CO_2 at 1000 K and 2 atm. total pressure. Assume $p_{CO}/p_{CO_2} = 10^3$.

    (ii) Suppose this gas mixture is brought in contact with zirconium metal, would ZrC form?

Solution   (i) For the reaction:

$$2CO\ (g) = CO_2\ (g) + C\ (s) \tag{E.10.5}$$

ΔG_5^0 can be obtained by combining the following reactions:

$$C(s) + \frac{1}{2} O_2(g) = CO(g); \qquad \Delta G_6^0 = -199.4 \times 10^3 \text{ J mol}^{-1} \tag{E.10.6}$$

$$C(s) + O_2(g) = CO_2(g); \qquad \Delta G_7^0 = -395.1 \times 10^3 \text{ J mol}^{-1} \tag{E.10.7}$$

The values of $\Delta G_6^0, \Delta G_7^0$ have been obtained from the data table in the Appendix. Hence,

$$\Delta G_5^0 = \Delta G_7^0 - 2\Delta G_6^0 = +3.7 \times 10^3 \text{ J mol}^{-1} = -RT \ln K_5$$

$$= -RT \ln \left(\frac{p_{CO_2} \times a_c}{p_{CO}^2} \right)_{eq} \tag{E.10.8}$$

Noting that $p_{CO}/p_{CO_2} = 10^3$, $p_{CO} \simeq 2$ atm, $p_{CO_2} \simeq \dfrac{2}{1000} = \dfrac{1}{500}$ atm., and substituting these values in Eq. (E.10.8), we get

a_C = activity of carbon at equilibrium of reaction (E.10.5) = 7.78×10^{-4}

$\mu_C = RT \ln a_C = -59.52$ kJ mol^{-1}

[*Note*: Actually, $RT \ln a_C = \mu_C - \mu_C^0$. However, by convention, μ_C^0 is zero [see Eq. (10.26)]. Also, as elucidated in section 10.1.3, although carbon is not present in the gas, there will be a finite value of μ_C in the gas.)

    (ii) For, $Zr(s) + C(s) = ZrC(s)$; $\Delta G_f^0 = -41.5$ kJ mol^{-1} at 1000 K   (E.10.9)

Again, $\Delta G_f^0 (ZrC) = -RT \ln \left(\dfrac{a_{ZrC}}{a_{Zr} \times a_C} \right)_{eq}$ (E.10.10)

Combining the above equations, and noting that the activities of ZrC and Zr are 1 since they are pure, we get

$$RT \ln(a_C)_{eq} = (\mu_C)_{\text{at eq with } Zr+ZrC} = -41.5 \text{ kJ mol}^{-1}$$

Since

$$(\mu_C)_{CO-CO_2 \text{ gas}} < (\mu_C)_{Zr+ZrC}, \text{ ZrC will not form.}$$

10.3 Thermodynamic Basis for Binary Alloy Phase Diagrams

Figure 10.1 shows the binary Pb-Sn phase diagram. In section 10.2.2, we have made some preliminary observations on these diagrams. Phase diagrams provide the basis for understanding the behaviour of alloys, as well as ceramic materials. Hence, they are very important in metallurgy and materials science, and these are typically taught in detail in separate courses. Therefore, the thermodynamic basis would be presented briefly for binary systems only, along with some examples. Ternary diagrams etc. will be omitted.

The terminology 'alloy' is traditionally reserved for metallic systems. However, all discussions are applicable to nonmetallic condensed systems as well. Therefore, for the sake of clarity the word 'alloy' is used in the heading of this section. Incidentally, nowadays, it is being used for nonmetallic materials also, e.g. *ceramic alloys*.

10.3.1 Free Energy vs. Composition Diagrams—General discussions

As elucidated in Section 5.7, any system tends to lower its Gibbs free energy at constant T, P. The equilibrium state corresponds to minimum $(G)_{T,P}$. Therefore, the nature of a phase diagram would depend on variation G with composition at various temperatures. As discussed several times earlier, the constant pressure restriction is not important.

Section 9.5 has dealt with thermodynamics of binary solutions. Figure 9.2 has schematically shown variation of ΔG^m, ΔH^m, ΔS^m with mole fraction for an ideal binary solution at constant temperature. Figure 9.6 has schematically shown variation of ΔG^m for regular binary solution for both positive and negative values of α.

For a regular binary solution A-B, where A and B are general symbols of the two components. From Eqs. (9.86) and (9.89), we have

$$\Delta G^m = \Delta G^{m,id} + G^{xs} = RT (X_A \ln X_A + X_B \ln X_B) + RT\alpha X_A X_B \qquad (10.33)$$

i.e.

$$\frac{\Delta G^m}{RT} = X_A \ln X_A + X_B \ln X_B + \alpha X_A X_B \qquad (10.34)$$

From Eq. (10.34), $\Delta G^m/RT$ as function of mole fraction can be calculated for different assumed values of α. This shows that, if $\alpha > 2.0$, the curve exhibits the maxima-minima behaviour. This is shown in Fig. 10.3(a). The curve is symmetric with respect to $X_B = 0.5$.

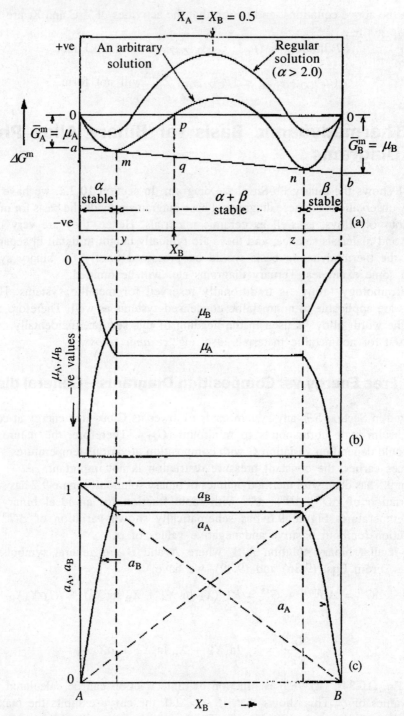

Fig. 10.3 Variation of: (a) free energy, (b) chemical potentials of A and B, and (c) activities of A and B, at constant temperature, for binary system A-B exhibiting maxima-minima type behaviour in ΔG^m (schematic).

Positive α means positive departure from Raoult's Law (see section 9.7.2). An arbitrary solution, which exhibits such positive departure but is not a regular solution, would not be symmetric with respect to $X_B = 0.5$. Figure 10.3(a) also shows such a behaviour schematically.

Let us now construct the common tangent *amnb* to the curve *mpn*. Let us consider an arbitrary composition, X_B. The point p is ΔG^m for the solution. Point q is ΔG^m for a mixture of two phases, whose compositions are fixed at m and n, respectively. Since q is lower than p, the two-phase mixture $\alpha + \beta$ would be stabler than the solution in this region.

The intercepts of the tangent at pure A and pure B provide the values of $\bar{G}_A^m$ and $\bar{G}_B^m$, respectively (see Section 9.5.3). Therefore,

$\bar{G}_A^m$ for α-phase at composition corresponding to $m = \bar{G}_A^m$ for β-phase at composition corresponding to n, i.e.

$$(\bar{G}_A^m)_{\alpha \text{ at } m} = (\bar{G}_A^m)_{\beta \text{ at } n} \tag{10.35}$$

Similarly,

$$(\bar{G}_B^m)_{\alpha \text{ at } m} = (\bar{G}_B^m)_{\beta \text{ at } n} \tag{10.36}$$

As discussed in Section 10.1.3, $\mu_A^0 = 0 = \mu_B^0$. Hence,

$$\bar{G}_A^m = \mu_A, \qquad \bar{G}_B^m = \mu_B \tag{10.37}$$

Combining Eqs. (10.35) to (10.37), we get

$$(\mu_A)_{\alpha \text{ at } m} = (\mu_A)_{\beta \text{ at } n} \tag{10.38}$$

$$(\mu_B)_{\alpha \text{ at } m} = (\mu_B)_{\beta \text{ at } n} \tag{10.39}$$

This is in conformity with the principle of equality of chemical potentials for phases at equilibrium (Section 10.1.2). Hence, it is confirmed that the two-phase mixture would have fixed compositions of α and β. It also follows that only α or β single phase solutions would be stable at the two-terminal regions since the ΔG^m curves are lower than straight lines joining any two points on the curve in the regions 'am' and 'nb'.

From Eqs. (10.38) and (10.39), μ_A and μ_B should be constant in two-phase mixture. Similarly, on the basis of Eq. (10.28), a_A and a_B also should be constant in the two-phase field $\alpha + \beta$. These are shown in Figures 10.3(b) and 10.3(c). It may be noted that $\mu_A = 0$ at $X_A = 1$, $\mu_B = 0$ at $X_B = 1$, as per convention already discussed in section 10.1.3.

10.3.2 Phase Diagrams with Miscibility Gap

The free energy vs. composition diagram of Fig. 10.3 schematically illustrates the thermodynamic basis regarding stabilities of the two-terminal solutions α and β and formation of $\alpha + \beta$ two-phase mixtures. There are phase diagrams which exhibit what is known as 'miscibility gap'. This is schematically illustrated in Fig. 10.4(a). Here, above the critical temperature T_{cr}, there is only one phase. But below T_{cr}, the single phase is transformed into two-phase mixture, which is the miscibility gap.

Assuming the solutions to be regular, ΔG^m is given by Eq. (10.33). Differentiating ΔG^m w.r.t. X_B, we obtain (noting $X_A = 1 - X_B$, and $dX_A = - dX_B$)

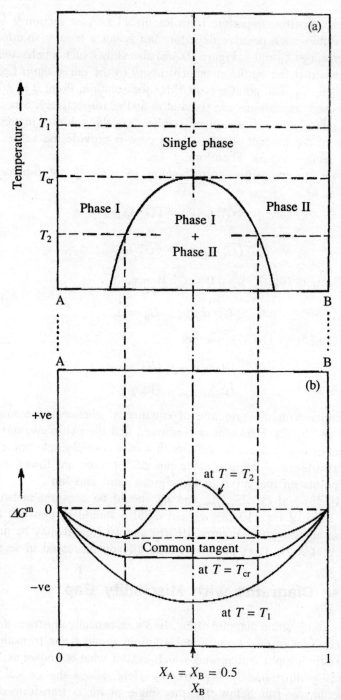

Fig. 10.4 (a) Phase diagram, and (b) corresponding free energy-composition diagrams for a binary system exhibiting miscibility gap, assuming regular solution behaviour (schematic).

$$\frac{\delta(\Delta G^{m})}{\delta X_{B}} = RT\left[\ln\frac{X_{B}}{X_{A}} + \alpha(X_{A} - X_{B})\right] \tag{10.40}$$

$$\frac{\delta^{2}(\Delta G^{m})}{\delta X_{B}^{2}} = RT\left(\frac{1}{X_{A}} + \frac{1}{X_{B}} - 2\alpha\right) \tag{10.41}$$

$$\frac{\delta^{3}(\Delta G^{m})}{\delta X_{B}^{3}} = RT\left(\frac{1}{X_{A}^{2}} - \frac{1}{X_{B}^{2}}\right) \tag{10.42}$$

From Eq. (10.42), we get $\dfrac{\delta^{3}(\Delta G^{m})}{\delta X_{B}^{3}} = 0$ at $X_{A} = X_{B} = 0.5$.

Since at $T = T_{cr}$, both $\dfrac{\delta^{3}(\Delta G^{m})}{\delta X_{B}^{3}}$ and $\dfrac{\delta^{2}(\Delta G^{m})}{\delta X_{B}^{2}}$ should be zero, the maxima of the miscibility gap would occur at $X_{B} = 0.5$, provided the solution is a regular one.

Putting $X_{A} = X_{B} = 0.5$ in Eq. (10.41), we get

$$\frac{\delta^{2}(\Delta G^{m})}{\delta X_{B}^{2}} = 0 \text{ at } 2\alpha = 4, \text{ i.e. } \alpha = 2 \tag{10.43}$$

Since $RT\alpha = \Omega$,

$$T_{cr} = \frac{\Omega}{2R} \tag{10.44}$$

for a binary regular solution.

From Eq. (9.89), we see that, since ΔH^{m} is approximately independent of temperature at a fixed composition, Ω may be taken as a constant, and not as a function of either temperature or composition for a particular binary regular solution. It has already been emphasized once, and concluded that

$$\alpha = \frac{\Omega}{RT} \propto \frac{1}{T}$$

which is the same as equation (9.92).

Thus, at $T > T_{cr}$, $\alpha < 2$, and the solution is single phase. On the other hand, at $T < T_{cr}$, $\alpha > 2$, and the miscibility gap appears. Figure 10.4(b) schematically shows the ΔG^{m} vs. X_{B} curves at different temperatures.

10.3.3 Free Energy vs. Composition Diagrams Involving Change of Standard State

In Figs. 10.3 and 10.4, in the free energy vs. composition curves, the standard states for both A and B were taken as the same (either both solid or both liquid). In actual phase diagrams,

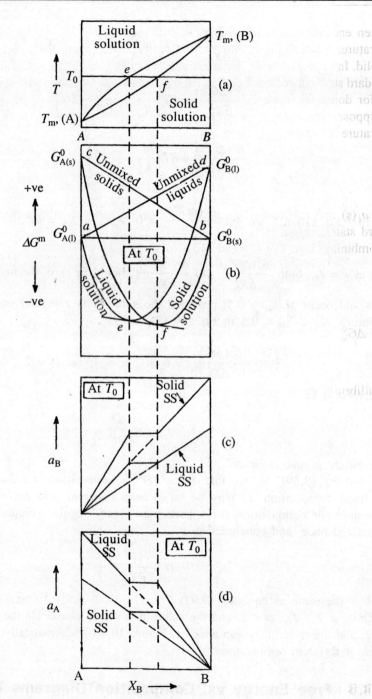

Fig. 10.5 (a) Phase diagram, (b) free energy-composition diagram at $T = T_0$, and (c), (d) activity-composition diagrams of B and A, respectively at $T = T_0$, for system with complete miscibility in liquid and solid phase (schematic).

we often encounter situations where one of them is liquid and the other solid at the same temperature. An example is Fig. 10.5(a), where at temperature T_0, pure A is liquid, but pure B is solid. In such a situation, for construction of free energy-composition diagrams, conversion of standard state either from liquid to solid or vice-versa would be required. The thermodynamic basis for doing the same is derived now.

Suppose free energy of pure solid i is G_i^0 (s) and that of pure liquid i is G_i^0 (l) at temperature T. Then,

$$RT \ln [a_i \text{ (s)}] = \bar{G}_i - G_i^0 \text{ (s)} \tag{10.45}$$

$$RT \ln [a_i \text{ (l)}] = \bar{G}_i - G_i^0 \text{ (l)} \tag{10.46}$$

where a_i(s) and a_i(l) refer to activity of i in the solution with respect to solid and liquid standard states, respectively.

Combining Eqs. (10.45) and (10.46), we get

$$RT \ln \left[\frac{a_i \text{(s)}}{a_i \text{(l)}} \right] = G_i^0 \text{(l)} - G_i^0 \text{(s)} = \Delta G_m^0 \tag{10.47}$$

where ΔG_m^0 is the free energy change of pure i upon melting at temperature T. Now,

$$\Delta G_m^0 = \Delta H_m^0 - T \Delta S_m^0 \tag{10.48}$$

At equilibrium melting temperature T_m,

$$\Delta G_m^0 = 0, \quad \Delta S_m^0 = \frac{\Delta H_m^0}{T_m} \tag{10.49}$$

Again,

$$\Delta G_m^0 \text{ (at T)} - \Delta G_m^0 \text{ (at } T_m) = \Delta G_m^0$$

$$= \Delta H_m^0 \text{ (at T)} - \Delta H_m^0 \text{ (at } T_m)$$

$$- T \Delta S_m^0 \text{ (at T)} + T_m \Delta S_m^0 \text{ (at } T_m) \tag{10.50}$$

i.e.

$$\Delta G_m^0 = \int_{T_m}^{T} [C_p \text{(l)} - C_p \text{(s)}] \, dT + T_m \Delta S_m^0 \text{ (at } T_m)$$

$$- T \left[\Delta S_m^0 \text{ (at } T_m) + \int_{T_m}^{T} \frac{[C_p \text{(l)} - C_p \text{(s)}]}{T} \, dT \right] \tag{10.51}$$

If it is assumed that $C_p(\text{l}) = C_p(\text{s})$, then Eq. (10.51) gets simplified into

$$\Delta G_m^0 \approx \Delta S_m^0 (\text{at } T_m) \cdot (T_m - T) \approx \frac{\Delta H_m^0 \text{ (at } T_m)}{T_m} \cdot (T_m - T) \tag{10.52}$$

i.e. ΔG_m^0 changes linearly with change of temperature.

EXAMPLE 10.7   Gold and silicon are mutually insoluble in the solid state and form a binary eutectic system with eutectic temperature of 636 K and composition of $X_{Si} = 0.186$ and $X_{Au} = 0.814$. Calculate the free energy of the eutectic melt relative to (i) unmixed (i.e. pure) solid Au and solid Si, and (ii) unmixed liquid Au and liquid Si. Assume C_P is the same for solid and liquid for both Au and Si.

Solution   (i) Free energy of the eutectic melt, relative to pure Au and pure Si solids at 636 K, is ΔG^m, where

$$\Delta G^m = X_{Au}\left[\bar{G}_{Au}(\text{eut}) - G^0_{Au}(s)\right] + X_{Si}\left[\bar{G}_{Si}(\text{eut}) - G^0_{Si}(s)\right]$$

$$= X_{Au}\left[\mu_{Au}(\text{eut}) - \mu^0_{Au}(s)\right] + X_{Si}\left[\mu_{Si}(\text{eut}) - \mu^0_{Si}(s)\right] \qquad \text{(E.10.11)}$$

Since the eutectic melt co-exists at equilibrium with both solid gold and solid silicon at 636 K,

$$\mu_{Au}(\text{eut}) = \mu^0_{Au}(s), \qquad \mu_{Si}(\text{eut}) = \mu^0_{Si}(s) \qquad \text{(E.10.12)}$$

Therefore, $\Delta G^m = 0$.

(ii) For free energy of the eutectic melt relative to pure liquid Au and pure liquid Si at 636 K,

$$\Delta G^m = X_{Au}\left[\bar{G}_{Au(\text{eut})} - G^0_{Au(l)}\right] + X_{Si}\left[\bar{G}_{Si(\text{eut})} - G^0_{Si(l)}\right] \qquad \text{(E.10.13)}$$

$$= X_{Au}\left[\bar{G}_{Au(\text{eut})} - G^0_{Au(s)}\right] + X_{Au}\left[\bar{G}^0_{Au(s)} - G^0_{Au(l)}\right]$$

$$+ X_{Si}\left[\bar{G}_{Si(\text{eut})} - G^0_{Si(s)}\right] + X_{Si}\left[\bar{G}^0_{Si(s)} - G^0_{Si(l)}\right]$$

Combining Eq. (E.10.13) with Eq. (E.10.12),

$$\Delta G^m = X_{Au}\left[G^0_{Au(s)} - G^0_{Au(l)}\right] + X_{Si}\left[G^0_{Si(s)} - G^0_{Si(l)}\right] \qquad \text{(E.10.14)}$$

Since C_P of solid and liquid are assumed as the same, from Eq. (10.52),

$$G^0(s) - G^0(l) = -\Delta G^0_m = -\frac{\Delta H^0_m(T_m)}{T_m}(T_m - 636) \qquad \text{(E.10.15)}$$

For Au,
$$T_m = 1336 \text{ K}, \quad \Delta H^0_m = 12{,}760 \text{ J mol}^{-1}$$
For Si,
$$T_m = 1683 \text{ K}, \quad \Delta H^0_m = 50{,}630 \text{ J mol}^{-1}$$

Substituting the values in Eqs. (E.10.15) and (E.10.14), we get

$$\Delta G_m = 0.814 \times \frac{12{,}760}{1336}(636 - 1336) + 0.186 \times \frac{50{,}630}{1683}(636 - 1683) = -11300 \text{ J mol}^{-1}.$$

10.4 Some Typical Free Energy vs. Composition Diagrams

10.4.1 Complete Solubility in Solid and Liquid Phases

Figure 10.5(a) shows the phase diagram schematically. An example is iron-nickel system. $T_m(A)$ and $T_m(B)$ are melting points of pure A and pure B, respectively. At $T = T_0$, pure A is liquid, but pure B is solid.

Figure 10.5(b) presents the free energy diagrams at T_0. It is to be noted that, in reality, we do not have either liquid or solid solution in the entire composition range. Hence, both curves are partly hypothetical and partly real at $T = T_0$. Since the stablest states of A and B are liquid and solid, respectively, $G_A^0(l)$ and $G_B^0(s)$ have been arbitrarily taken as zero. $G_A^0(s)$ and $G_B^0(l)$ are values at metastable states. Hence, they correspond to hypothetical standard states of A and B, and the values can be calculated by employing Eqs. (10.47) and (10.52); *ef* is the common tangent.

Figures 10.5(c) and 10.5(d) show activity vs. composition relations for B and A, respectively at T_0. Curves corresponding to liquid solution have liquid A and B have standard states, whereas curves corresponding to those of solid solutions have solid standard states.

10.4.2 Eutectic Phase Diagram with Finite Solid Solubilities

Figures 10.6(a) and 10.7(a) are the schematic phase diagrams. The terminal solid solutions are α and β. Figures 10.6(b) and 10.6(c) present free energy vs. composition and activity vs. composition diagrams at $T = T_1$. Figures 10.7(b) and 10.7(c) are those at $T = T_2$. Curves in both Figs. 10.6(b) and 10.7(b) for the liquid solution have liquid standard states (real or hypothetical, as the case may be). Curves for α and β solid solutions, respectively, have solid A and B as standard states. The activity vs. composition curves have employed the real (i.e. stablest) standard states for each component.

10.4.3 Phase Diagram with an Intermediate Phase

Figure 10.8(a) shows the nature of free-energy composition curves at temperature T_1. Figure 10.8(b) is a schematic of a phase diagram with an intermediate phase; α and γ are the two terminal solid solutions. In addition, a high-melting intermediate solid phase β is present.

10.5 Solubility of Solute in a Phase in Binary System

The solubility of a component in a solution refers to the concentration of the same in the solution co-existing with another phase at equilibrium. In this section, this will be illustrated and some quantitative relations derived using simple binary system A-B.

Consider Fig. 10.7(a). The solubilities of component B at $T = T_2$ are $X_B^\alpha(\text{sat})$ and $X_B^\beta(\text{sat})$

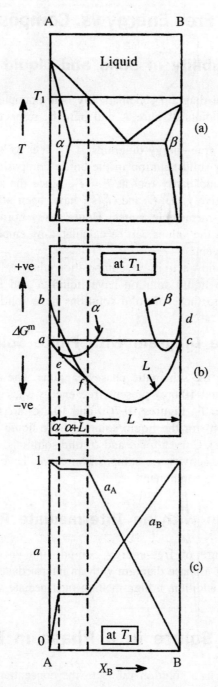

Fig. 10.6 (a) Phase diagram, (b) free energy-composition diagram at $T = T_1$, and (c) activity-composition diagram at $T = T_1$, for a binary eutectic system with terminal solid solubilities (schematic).

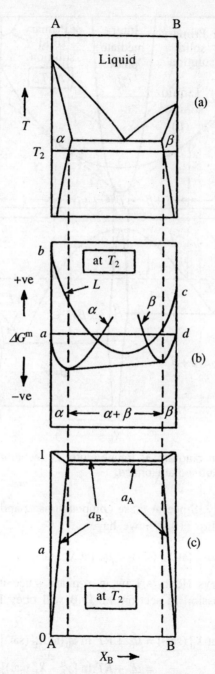

Fig. 10.7 (a) Phase diagram, (b) free energy-composition diagram at $T = T_2$, and (c) activity-composition diagram at $T = T_2$, for a binary eutectic system with terminal solid solubilities (schematic).

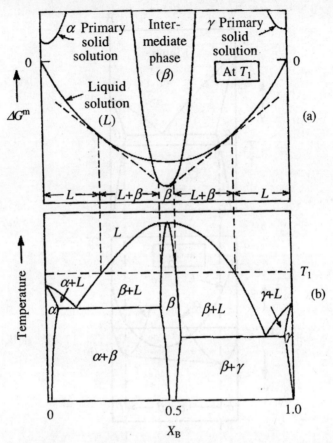

Fig. 10.8 (a) ΔG^m-composition diagram at temperature T_1 (schematic), for (b) binary phase diagram with an intermediate phase.

in α and β phases, respectively. Since at these compositions α and β phases are co-existing at equilibrium, according to Eq. (10.19), we have

$$\mu_B \text{ [at } X_B^\alpha \text{ (sat)]} = \mu_B \text{ [at } X_B^\beta \text{ (sat)]} \qquad (10.53)$$

Let us assume that B obeys Henry's Law in α-phase since its concentration is small. Therefore, based on the discussion in section 9.5.1, B will obey Raoult's Law in β-phase. Hence,

$$\mu_B^\alpha \text{ [at } X_B^\alpha \text{ (sat)]} = \mu_B^0 + RT \ln a_B^\alpha [\text{at } X_B^\alpha \text{ (sat)}]$$

$$= \mu_B^0 + RT \ln [\gamma_B^0 \cdot X_B^\alpha \text{ (sat)}]$$

$$= \mu_B^0 + RT \ln \gamma_B^0 + RT \ln [X_B^\alpha \text{ (sat)}] \qquad (10.54)$$

$$\mu_B^\beta \text{ [at } X_B^\beta \text{ (sat)]} = \mu_B^0 + RT \ln a_B^\beta [\text{at } X_B^\beta \text{ (sat)}]$$

$$= \mu_B^0 + RT \ln [X_B^\beta \text{ (sat)}] \qquad (10.55)$$

Equating Eqs. (10.54) and (10.55), we get

$$\ln \left\lfloor X_B^\beta (\text{sat}) \right\rfloor = \ln \left\lfloor X_B^\alpha (\text{sat}) \right\rfloor + \ln \gamma_B^0 \qquad (10.56)$$

i.e.

$$\gamma_B^0 = \frac{X_B^\beta (\text{sat})}{X_B^\alpha (\text{sat})} \qquad (10.57)$$

Suppose now that there is no solid solubility in β-phase, as shown in Fig. 10.9. Then $X_B^\beta (\text{sat}) = 1$. Hence,

$$X_B^\alpha (\text{sat}) = \frac{1}{\gamma_B^0} \qquad (10.58)$$

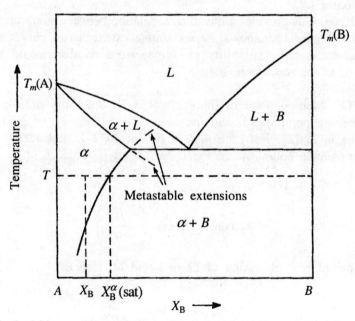

Fig. 10.9 Hypothetical binary phase diagram illustrating derivation of solubility equation for B in α-phase; dashed lines show metastable extensions.

Let us consider a composition X_B in α-phase, and let a_B be the activity of B at X_B. Then,

$$a_B = \gamma_B^0 X_B = \frac{X_B}{X_B^\alpha (\text{sat})} \qquad (10.59)$$

From Eq. (10.25),

$$\mu_B - \mu_B^0 = RT \ln a_B = RT \left\lfloor \ln X_B - X_B^\alpha (\text{sat}) \right\rfloor \qquad (10.60)$$

At fixed composition X_B, on the basis of Gibbs–Helmholtz equation i.e. Eqs. (6.28), (6.29) and their applications to partial molar properties [Eqs. (9.31), (9.33)]

$$\frac{d\left[(\mu_B - \mu_B^0)/T\right]}{d(1/T)} = \bar{H}_B - H_B^0 = \bar{H}_B^m = -R\left[\frac{d\ln[X_B^\alpha(\text{sat})]}{d(1/T)}\right] \quad (10.61)$$

Again, from Eq. (10.60), $\dfrac{d\left[(\mu_B - \mu_B^0)/T\right]}{d(1/T)} = -R\left[\dfrac{d\ln[X_B^\alpha(\text{sat})]}{d(1/T)}\right]$ (10.62)

Combining Eqs. (10.61) and (10.62), and integrating

$$X_B^\alpha(\text{sat}) = A\exp\left(-\frac{\bar{H}_B^m}{RT}\right) \quad (10.63)$$

where A is a constant.

Equation (10.63) thus provides a quantitative relation between solubility of B in α-phase with temperature. Figure 10.9 shows the *metastable extensions* of curves by dashed lines. It should be noted that the solubility vs. temperature relation would be governed by Eq. (10.62) even in the metastable state.

EXAMPLE 10.8 Melting point of lithium is 453 K. Solubility of iron in liquid Li is minimum at melting point of Li, and increases with increasing temperature. Given that solubilities of Fe in Li [X_{Fe} (sat)] are 4.36×10^{-4} at 1473 K and 4.97×10^{-6} at 673 K, determine the minimum solubility.

Solution   From Eq. (10.63),

$$X_{Fe}(\text{sat}) = A\exp\left(-\frac{\bar{H}_{Fe}^m}{RT}\right) \quad (E.10.16)$$

where $\bar{H}_{Fe}^m$ = partial heat of mixing of Fe in liquid Li. Hence,

$$\ln[X_{Fe}(\text{sat})] = \ln A - \frac{\bar{H}_{Fe}^m}{RT} \quad (E.10.17)$$

Putting in the values of X_{Fe} (sat) at 1473 K and 673 K in Eq. (E.10.17), we obtain

$$-7.738 = \ln A - \frac{\bar{H}_{Fe}^m}{R \times 1473} \quad (E.10.18)$$

$$-12.212 = \ln A - \frac{\bar{H}_{Fe}^m}{R \times 673} \quad (E.10.19)$$

Solving the above equations, $\ln A = -3.97$, $\bar{H}_{Fe}^m = 46143$ J mol^{-1}. Putting the values of the above equations, X_{Fe} (sat) at 453 K = 9×10^{-8}.

[*Note:* A may be considered as equal to $\exp(\bar{S}_{Fe}^m/R)$ which gives $\bar{S}_{Fe}^m = -40.65$ J mol^{-1} K^{-1}]

10.5.1 Solubility of Metastable Phases

Figure 10.9 shows that solubilities of B in α-phase in metastable extensions are larger than those in the stable regions of the bounding curves. The schematic free-energy composition diagram for the iron-carbon system at a temperature above the eutectoid temperature is shown in Fig. 10.10. Here austenite is the stable phase at the Fe-end. For carbon, graphite is the stable and cementite is the metastable state. This means that the chemical potential of carbon in cementite is higher than that in graphite. As Fig.10.10 shows, this leads to a higher solubility of carbon in austenite in equilibrium with cementite as compared to that in equilibrium with graphite. Experimental phase diagram data have confirmed this.

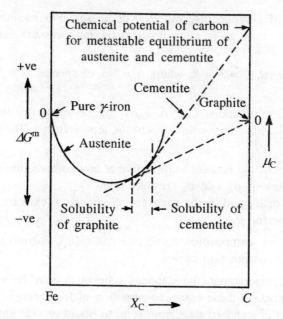

Fig. 10.10 Schematic diagram of Fe-C system explaining why solubility of carbon in austenite is higher at equilibrium with cementite as compared to that for graphite.

10.6 Summary

1. There are several definitions of chemical potential (μ_i) of a component i in a solution. However, the commonly employed definition in chemical and metallurgical thermodynamics is in terms of Gibbs free energy, as follows:

$$\mu_i = \left(\frac{\delta G'}{\delta n_i}\right)_{P,T,\,n_1,\,n_2,\,...,\,\text{except } n_i}$$

This way, μ_i becomes the same as the partial molar free energy of component i ($\bar{G}_i$).

2. For phases at equilibrium at constant T and P, the chemical potential of a component is the same in all phases. If they are unequal, then the component may be transferred from higher to lower chemical potential.

3. By convention, the chemical potential of a component at its standard state is assumed to be zero. Hence,

$$\mu_i - \mu_i^0 = \mu_i = \bar{G}_i^m = RT \ln a_i$$

4. Gibbs Phase Rule has been derived as:

$$\text{Degree of freedom } (F) = C - P + 2$$

where C = No. of components, and P = of phases. For equilibria of nonreacting condensed phases, such as in alloy phase diagrams, pressure has negligible effect. Hence, $F = C - P + 1$.

For reacting systems, $C = N - R$, where N = No. of species, R = No. of independent reactions.

5. Binary alloy phase diagrams (say $A - B$) can be theoretically constructed from variation of ΔG^m with composition at various temperatures. The basis for the same is:

   (a) intersections of the tangent to the curve at any composition at pure A and pure B give values of μ_A and μ_B, respectively.
   (b) The values of μ_A will be the same for all phases co-existing at equilibrium; the same is true for μ_B.

6. Thus, from ΔG^m vs. composition curve, we can also construct μ_A and μ_B, as well as a_A and a_B vs. composition curves.

7. Often, at some temperatures, the standard state of A may be solid and that of B liquid, or vice versa. In these cases, construction of free energy composition curves requires changing of standard state from solid to liquid or vice versa. Equations have been derived for the same in this chapter.

8. The procedure for construction of phase diagrams etc. from ΔG^m vs. composition curves has been illustrated for some typical phase diagrams.

9. Finally, the solubility of a solute in a phase in binary system has been derived as function of temperature.

PROBLEMS

10.1 Calculate the chemical potential of carbon in a gas mixture containing 30% CO, 20% CO_2, 50% N_2 at 1300 K. Will this gas saturate a piece of steel with carbon? The total pressure is 5 atm.

10.2 Pure solids of Si, SiO_2 and Si_3N_4 are equilibrated with a $O_2 + N_2$ gas mixture at 1000 K. How many degrees of freedom does this equilibrium have? What is the value of chemical potential of N_2 at equilibrium? While maintaining the gas composition constant, if the temperature is raised to 1200 K, what happens?

10.3 How many degrees of freedom does the system consisting of FeO-MnO solution + Fe-Mn solution + O_2-N_2 gas mixture have?

10.4 Calculate the chemical potential of S_2 in a gas mixture consisting of 2% H_2S, 50% H_2, and the rest N_2 at 1200 K. From chemical potential considerations, will this mixture convert metallic copper into Cu_2S?

10.5 Calculate the chemical potential of oxygen gas (i.e. O_2) in a system consisting of pure liquid iron and a liquid slag containing FeO at equilibrium at 1600°C; the activity of FeO in the slag is given as 0.45.

10.6 Derive an expression for the free energy of fusion of iron (ΔG_m) as function of temperature. Assume that C_P of liquid Fe is larger than that of solid Fe by 1.3 J mol^{-1} K^{-1}. Heat of fusion of Fe is 15,360 J mol^{-1} at its melting point of 1535°C.

10.7 The solubility of carbon in liquid aluminium is 6 ppm at 960°C and 12.5 ppm at 1000°C.

   (i) Calculate the solubility at melting point of Al, i.e. at 660°C. Assume dilute solution behaviour.

   (ii) Predict whether solid aluminium carbide (Al_4C_3) will form at carbon saturation of liquid Al at 660°C. It is given that, for the reaction:

$$4Al(l) + 3C(s) = Al_4C_3(s), \qquad \Delta G^0 = -266,521 + 96.23\ T\ \text{J mol}^{-1}$$

10.8 The Ag-Cu alloy has a eutectic phase diagram with eutectic temperature of 780°C. Below the eutectic temperature, the eutectic liquid decomposes into a mixture of α(Cu-rich) and β(Ag-rich) solid solutions. At 500°C, the saturated α- and β-phases respectively contain 98.8 and 3.35 atom % Cu. Assuming Henrian behaviour of Cu in β-phase and Raoultian behaviour in the α-phase, calculate Henry's Law constant of Cu in β-phase.

Reaction Equilibria Involving Condensed Phases with Variable Compositions

In Chapter 8, we presented techniques of calculations of reaction equilibria involving ideal gases and pure condensed phases. These were illustrated with worked-out examples as well. This chapter is concerned with equilibrium calculations where the solid and liquid are not necessarily pure, but are solutions. However, before doing that, we shall give some examples of construction and/or use of phase diagrams for reactive systems.

In section 8.3.3, the phase stability diagram of Ni-S-O system has been derived on the basis of gas-solid reaction equilibria. There, all solid phases were taken as pure, and hence, activities of solids are 1. In this chapter, first we shall give some further examples of pure solids.

11.1 Analysis of Phase Stabilities of Reactive Stoichiometric Compounds—The Si-C-O system

In the Si-C-O system at temperatures of interest, there are four stable condensed phases—silicon, carbon, silica, and silicon carbide. C and SiC remain solids at all temperatures of interest in manufacture of SiC, and may be assumed to be pure. Si melts at 1683 K. It dissolves some carbon whose concentration is very low, and liquid Si may be assumed to be pure. SiO_2 melts at 2001 K and remains essentially a stoichiometric compound.

Besides the condensed phases, there would be a gas phase containing CO, CO_2 and SiO at high temperature. The following four reactions may be considered:

$$C(s) + CO_2(g) = 2CO(g) \qquad (11.1)$$

$$SiO_2(s, l) + CO(g) = SiO(g) + CO_2(g)^* \tag{11.2}$$

$$SiC(s) + 2CO_2(g) = SiO(g) + 3CO(g) \tag{11.3}$$

$$Si(s, l) + CO_2(g) = SiO(g) + CO(g) \tag{11.4}$$

The above reactions are all independent since none of them can be arrived at by combining the others. Section 10.2 has discussed phase rule and its application to reaction systems as:

$$\text{Degree of freedom } F = C - P + 2$$

$$C = N - R$$

which are the same as Eqs. (10.31) and (10.32). In this example,

$$N = 7 \ (Si, SiO_2, SiC, C, CO, CO_2, SiO)$$

$$R = \text{No. of independent reactions} = 4$$

Hence, if P and T are fixed, then the maximum number of phases co-existing at equilibrium will be 3 (i.e. for $F = 0$).

The relations for equilibrium constants for the above reactions are

$$K_1 = \frac{(p_{CO})^2}{p_{CO_2}}, \quad K_2 = \frac{p_{CO_2} \cdot p_{SiO}}{p_{CO}}$$

$$K_3 = (p_{CO})^3 \, p_{SiO}/(p_{CO_2})^2, \quad K_4 = \frac{p_{CO} p_{SiO}}{p_{CO_2}} \tag{11.5}$$

where p refers to partial pressures at equilibrium. Again,

$$p_{CO} + p_{CO_2} + p_{SiO} = P_T = \text{total pressure} \tag{11.6}$$

Figure 11.1 presents calculated $\log p_{CO_2}$ vs. $\log p_{SiO}$ curves at equilibria with reactions (11.1)–(11.4), at 1700 K and $P = 1$ atm on the basis of Eqs. (11.5) and (11.6) (Ref.: A. Ghosh and G.R. St. Pierre, *Trans. Met. Soc. AIME*, 245, 1969, p. 2106), and from the standard free energies of formation of compounds.

Derivation of the ternary phase diagram from Fig. 11.1 is based on the following principles:

1. Phase rule allows equilibrium co-existence of a maximum of three phases at fixed P and T.
2. The gas compositions in equilibrium with single condensed phases are given by the curves in Fig. 11.1.
3. At constant p_{SiO}, SiO_2 is unstable with respect to a gas composition which contains less CO_2 than that corresponding to the SiO_2–curve. On the other hand, at a constant p_{SiO}, C, Si and SiC are unstable if the gas contains more CO_2 than those corresponding to their equilibrium curves. These statements follow directly from Eqs. (11.1)–(11.4).

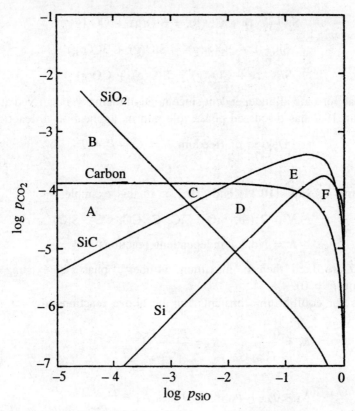

Fig. 11.1 Log p_{CO_2} vs. log p_{SiO} in the gas phase in equilibrium with various condensed phases at 1700 K, and at $P_T = 1$ atm. [A. Ghosh and G.R. St. Pierre, *Trans. AIME*, **245**, 2106, (1969.)]

Using the above principles, all regions (A, B, C etc.) and all points of intersection in Fig. 11.1 were examined to find out the stable co-existing phases in the system. Thus we arrive at Table 11.1.

Table 11.1 Derivation of Phase Fields

Phase Fields (Fig.11.2)	Phase field derived from Fig. 11.1
Gas	Region B
SiO_2 + gas	SiO_2-curve
C + gas	Carbon-curve
C + SiO_2 + gas	Carbon-SiO_2 Intersection
C + SiO_2	Region C
SiC + C + SiO_2	Region D
SiC + SiO_2	Region E
Si + SiC + SiO_2	Region F

As an explanation of Table 11.1, let us consider a gas composition in region C of Fig. 11.1. Carbon will be stable, since P_{CO_2} is below that at equilibrium of reaction (11.1), meaning that the reaction will tend to proceed from RHS to LHS, thus depositing carbon. SiO_2 will be stable since P_{CO_2} in region C is above that of equilibrium for SiO_2-gas. Thus the reaction (11.2) shall proceed from RHS to LHS. However, SiC and Si will not be stable if similar arguments are employed for their reactions.

Figure 11.2 presents the calculated Si-C-O ternary phase diagram based on the stable phases of Table 11.1.

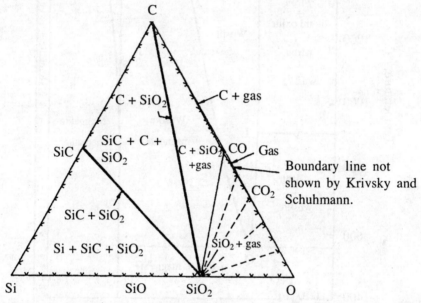

Fig.11.2 Si-C-O phase diagram at 1700 K and 1 atm [A. Ghosh and G.R. St. Pierre, *Trans. AIME*, **245**, 2106 (1969)].

11.2 Phase Diagram with Reactive Non-stoichiometric Compounds—The iron-oxygen system

11.2.1 Phase Diagram

Oxides of iron, viz. Magnetite (Fe_3O_4) and Hematite (Fe_2O_3), are minerals from which iron is extracted. They have many other industrial uses also. The phase diagram of iron-oxygen system is presented in Fig. 11.3. Its special features are as follows:

- There are three solid oxides, viz. FeO (Wustite), Fe_3O_4 (Magnetite), and Fe_2O_3 (Hematite).
- Wustite is a non-stoichiometric compound. Hence it is designated either as "FeO" or Fe_xO, where x ranges between the overall limits of 0.835–0.945. Moreover, it is stable only above 570°C.

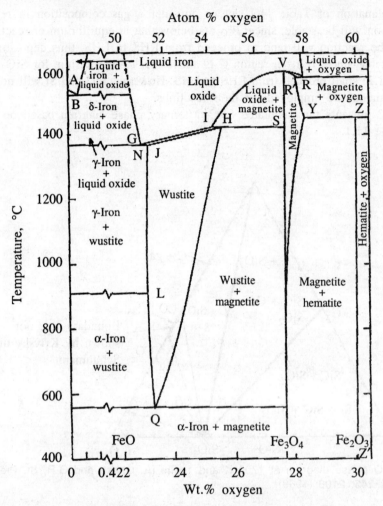

Fig.11.3 Iron-oxygen phase diagram [L.S. Darken and R.W. Gurry, *J.Am. Chem. Soc.*, **68**, 798 (1946)].

- The oxygen-to-iron ratio in the system at a fixed temperature depends on partial pressure (i.e. chemical potential) of oxygen in the gas phase with which the solid phases are brought to equilibrium.

There are several other phase diagrams of oxide system etc., where such features are present, i.e. non-stoichiometric compounds and dependence of composition and nature of phases of the condensed system on gas composition. But the Fe-O system constitutes the most common one.

11.2.2 Activities in the Wustite Phase

Wustite phase may be considered as a solid solution of FeO and Fe_3O_4. It is a ionic solid having Fe^{2+}, Fe^{3+} and O^{2-} ions. There is appreciable concentration of vacancies in the cationic lattice due to Fe deficiency. Therefore, the activity of FeO would continuously vary in the Wustite phase field. The choice of standard state poses problem since stoichiometric FeO is not available. Hence, it is appropriate to talk of an "FeO", where "FeO" refers to non-stoichiometric Wustite. *By convention, "FeO" co-existing with metallic* Fe *at equilibrium is taken as its standard state* (e.g. point '*I.*') at any temperature, *T*.

The choice of standard state for ions poses problem. Let us take the example of aqueous solution and consider the case of $CuSO_4$ dissolved in water. The standard state of Cu^{2+} and SO_4^{2-} are unknown since they are not available in pure form. The difficulty is overcome by assuming

$$a_{Cu}^{2+} \cdot a_{SO_4}^{2-} = a_{CuSO_4} \, (aq) \tag{11.7}$$

If the solution is *saturated with pure solid* $CuSO_4$, then as per accepted convention,

$$a_{Cu}^{2+} \cdot a_{SO_4}^{2-} = 1 \tag{11.8}$$

With similar approach, it may be written that in Wustite phase,

$$a_{Fe} \cdot a_O = a_{FeO} \tag{11.9}$$

At a temperature *T* and the corresponding point on the line QLJ, $a_{FeO} = 1$ as assumed earlier. Since a_{FeO} is at equilibrium with pure Fe, $a_{Fe} = 1$. Hence, from Eq. (11.9), $a_O = 1$ in Wustite co-existing with metallic iron.

The reaction of oxygen may be written as

$$(O)_w = \frac{1}{2} O_2 \, (g) \tag{11.10}$$

where the subscript 'w' denotes Wustite, for which

$$K_{10} = \left[\frac{(p_{O_2})^{1/2}}{(a_0)_w} \right]_{eq} \tag{11.11}$$

Combining Eq. (11.11) with equation at standard state, where $a_o = 1$, we obtain

$$(a_0)_w = \left[\frac{p_{O_2}}{p_{O_2}^0} \right]^{1/2} \tag{11.12}$$

where $p_{O_2}^0$ represents p_{O_2} in equilibrium with Fe + FeO.

The procedure adopted for experimental determination of a_O, a_{Fe}, a_{FeO} in Wustite field is as follows:

1. Temperature and p_{O_2} of gas were fixed. The solids were allowed to come to equilibrium with the gas. Then the solid sample was rapidly quenched to room temperature, and analyzed for Fe and O content.

2. The solid phases were identified by X-ray diffraction studies.
3. a_O in Wustite field was calculated at various compositions from Eq. (11.12).
4. a_{Fe} was calculated by the Gibbs–Duhem integration on the basis of Eq. (9.62) from a_O vs. composition data.
5. a_{FeO} was calculated from the above using Eq. (11.9).

Figure 11.4 presents activity vs. composition data in the Fe-O system at 1100°C. It may be noted that the minimum value of a_O is 1. This is due to the procedure adopted.

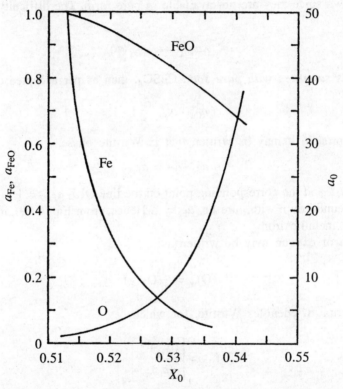

Fig. 11.4 Activities of iron, oxygen, and Wustite in the Wustite phase at 1100°C.

11.3 Reaction Equilibria Involving Solutions of Condensed Phases

The thermodynamic behaviour of solutions has been discussed in Chapter 9. Now we shall present some examples of reaction equilibria calculations involving solutions of condensed phases. However, first we shall discuss the general considerations.

11.3.1 General Considerations

For this, let us reconsider oxidation of Mn, viz.

$$2Mn(s) + O_2(g) = 2MnO(s)$$

which is the same as Eq. (8.18).

In section 8.3.2, both Mn and MnO were assumed to be pure; that is a special case. The general situation is that Mn is present in metallic solution, and MnO is present as an oxide solution in association with other oxides. The solutions may be solid or liquid. By convention, we write the above reaction as

$$2[Mn] + O_2(g) = 2(MnO) \tag{11.13}$$

or

$$2\underline{Mn} + O_2(g) = 2(MnO) \tag{11.14}$$

The round brackets denotes a compound present in a non-metallic solution (oxide, sulphide, chloride etc.), while the square bracket or underline denotes an element in a metallic solution. It should be mentioned here that all components are present as elements in a metallic solution. In other words, *it is an atomic solution*. Therefore, solutions containing compounds will exist as separate phases.

The equilibrium constant for the above reaction is

$$K_{13} = \left\{ \frac{(a_{MnO})^2}{[a_{Mn}]^2 \, p_{O_2}} \right\}_{eq} \tag{11.15}$$

For pure Mn and pure MnO, if $p_{O_2}^0$ is equilibrium p_{O_2}, then

$$K_{13} = \left(\frac{1}{p_{O_2}^0} \right)_{eq} \tag{11.16}$$

From Eqs. (11.15) and (11.16),

$$p_{O_2} = p_{O_2}^0 \frac{(a_{MnO})^2}{[a_{Mn}]^2} \tag{11.17}$$

From the Ellingham diagram (Fig. 8.2), at 1200 K, $p_{O_2}^0 = 5 \times 10^{-20}$ atm. Let $[a_{Mn}]$ = 0.3 in a Mn-Fe solid solution and $(a_{MnO}) = 0.45$ in a MnO-SiO$_2$ solid solution. Then, p_{O_2} at equilibrium with these will be 11.25×10^{-20} atm.

A major part of thermodynamic measurements on metals, alloys and other materials consists of experimental determination of activities of components in various solid and liquid solutions at high temperature. With extensive measurements by physical chemists, metallurgists and materials scientists over a period of 60 to 70 years, activity versus composition data are available in most solutions of interest.

EXAMPLE 11.1 Consider the reaction:

$$[Fe] + H_2O(g) = FeO(s) + H_2(g), \text{ at } 910°C \text{ and } 1 \text{ atm. total pressure.}$$

When the iron is pure, the gas mixture at equilibrium contains 60.2% H_2 and 39.8% H_2O(by volume, as per standard convention for gases). If Fe is present as a solid solution of Fe and Ni($X_{Fe} = 0.721$), the equilibrium gas composition is 51.9% H_2, and 48.1%H_2O.

Calculate the activity of Fe in the alloy, assuming FeO to be pure.

Solution The equilibrium constant for the above reaction is

$$K = \left(\frac{a_{FeO} \times p_{H_2}}{a_{Fe} \times p_{H_2O}} \right)_{eq} \tag{E.11.1}$$

Since FeO is pure, $a_{FeO} = 1$. Hence the above equation may be rewritten as

$$K = \left(\frac{p_{H_2}}{p_{H_2O}} \right)_{eq} = \frac{0.602}{0.398} \qquad \text{for pure Fe}$$

$$= \left(\frac{p_{H_2}}{p_{H_2O}} \frac{1}{a_{Fe}} \right)_{eq} = \frac{0.519}{0.481} \frac{1}{a_{Fe}} \qquad \text{for Fe–Ni} \tag{E.11.2}$$

From Eq. (E.11.2), a_{Fe} in the alloy = 0.713. Since $X_{Fe} = 0.721$, it is an ideal solution (within error limit).

EXAMPLE 11.2 The p_{H_2}/p_{H_2O} ratio in an H_2-H_2O gas mixture in equilibrium with pure liquid Pb and lead silicate melt, at $X_{PbO} = 0.7$, is 5.66×10^{-4} at 900°C. The corresponding value at 1100°C is 1.2×10^{-3}. Calculate the partial molar heat of mixing of liquid PbO in this lead silicate melt.

Solution   900°C = 1173 K, 1100°C = 1373 K

The reaction is:

$$Pb(l) + H_2O(g) = (PbO)_{in \text{ melt}} + H_2(g) \tag{E.11.3}$$

for which

$$\Delta G_3^0 = -RT \ln K_3 = -RT \ln \left(\frac{a_{PbO} \times p_{H_2}}{a_{Pb} \times p_{H_2O}} \right) \tag{E.11.4}$$

Again,

$$\Delta G_3^0 = \Delta G_f^0(PbO) - \Delta G_f^0(H_2O) = (-191,750 + 79.1T) - (-246,000 + 54.8T)$$

$$= 54,250 + 24.3T \text{ J mol}^{-1}$$

Hence,

$$\Delta G_3^0 = 82{,}754 \text{ J}, \qquad K_3 = 2.06 \times 10^{-4} \text{ at } 1173 \text{ K},$$

$$\Delta G_3^0 = 87{,}614 \text{ J}, \qquad K_3 = 4.64 \times 10^{-4} \text{ at } 1373 \text{ K},$$

Noting that $a_{Pb} = 1$,

$$a_{PbO} = K\left(\frac{p_{H_2O}}{p_{H_2}}\right) \tag{E.11.5}$$

Putting in the values given in Eq. (E.11.5), we get $a_{PbO} = 0.364$ and 0.387, respectively at 1173 K and 1373 K.

Now,

$$\bar{G}_{PbO}^m = RT \ln(a_{PbO}) \tag{E.11.6}$$

Hence,

$\bar{G}_{PbO}^m = -9856$ and $-10{,}837$ J mol^{-1}, respectively at 1173 K and 1373 K.

Partial molar heat of mixing of liquid PbO in the lead silicate melt ($\bar{H}_{PbO}^m$) is related to $\bar{G}_{PbO}^m$ by Gibbs–Helmholtz equation as

$$\left[\frac{\delta\left(\dfrac{\bar{G}_{PbO}^m}{T}\right)}{\delta\left(\dfrac{1}{T}\right)}\right]_{P,\,comp} = \bar{H}_{PbO}^m \tag{E.11.7}$$

Assuming $\bar{H}_{PbO}^m$ to be independent of temperature, integration of Eq. (E.11.7) leads to

$$\bar{H}_{PbO}^m = \frac{\left(\dfrac{\bar{G}_{PbO}^m}{T}\right)_{1373} - \left(\dfrac{\bar{G}_{PbO}^m}{T}\right)_{1173}}{\left(\dfrac{1}{1373} - \dfrac{1}{1173}\right)} \tag{E.11.8}$$

Putting in values, we get $\bar{H}_{PbO}^m = -4803$ J mol^{-1}.

EXAMPLE 11.3 Calculate the partial pressure of nitrogen gas over an eutectic liquid solution of Al-Si alloy at the eutectic temperature, at equilibrium with solid Si_3N_4 (Johnson and Stracher, 1995, p. 134).

Solution The relevant reaction is

$$3[Si]_{\text{Al-Si melt}} + 2N_2(g) = Si_3N_4(s) \tag{E.11.9}$$

for which

$$\Delta G_9^0 = -RT \ln K_9 = -RT \ln \left\{ \frac{a_{Si_3N_4}}{[a_{Si}]^3 \times p_{N_2}^2} \right\}_{eq} \qquad \text{(E.11.10)}$$

Since Si_3N_4 is pure, the activity of Si_3N_4 is 1.

From Al-Si phase diagram, the eutectic temperature = 850 K, and at eutectic point, $X_{Si} = 0.122$. At this temperature, pure Si is a solid. But for application of Eq. (E.11.10), liquid Si is to be employed as the standard state, and ΔG_9^0 is to have the appropriate value for this.

ΔG_9^0 can be calculated from the following data obtained from literature sources.

$$\Delta G_{11}^0 = -753,190 + 336.43T \text{ J mol}^{-1} \text{ for } 3Si(s) + 2N_2(g) = Si_3N_4(s) \quad \text{(E.11.11)}$$

$$\Delta G_m^0 = 50,630 - 29.91T \text{ J mol}^{-1} \text{ for } Si(s) = Si(l) \qquad \text{(E.11.12)}$$

From Hess' Law,

$$\Delta G_9^0 = \Delta G_{11}^0 - 3\Delta G_m^0 = (-753,190 + 336.43 \times 850)$$

$$-3(50,630 - 29.91 \times 850) = -542,844 \text{ J mol}^{-1}$$

Solving Eq. (E.11.10), we get

$$\{[a_{Si}]^3 \times p_{N_2}^2\}_{eq} = 4.36 \times 10^{-34}$$

From literature sources, at 850 K and $X_{Si} = 0.122$ in liquid Al-Si solution, with reference to liquid Si as standard state,

$$\bar{G}_{Si}^m = RT \ln a_{Si} = -24,324 \text{ J}$$

This yields

$$a_{Si} = 3.28 \times 10^{-5}$$

Solving, p_{N_2} at equilibrium with Al-Si melt and $Si_3N_4(s) = 3.7 \times 10^{-15}$ atm.

If formation of Si_3N_4 is not desired, then p_{N_2} in the atomosphere should be kept below this value. [*Note:* If solid Si is used as standard state, then a_{Si} in Al–Si eutectic with reference to solid Si is to be emloyed.]

EXAMPLE 11.4 The reaction:

$$2[Cu]_{\text{in liq. Pb}} + PbS(s) = Cu_2S(s) + Pb(l) \qquad \text{(E.11.13)}$$

allows removal of Cu present as impurity in molten lead. The solid sulphides are present as pure compounds, and Pb is insoluble in solid Cu. The solubility of Cu in liquid Pb is given as

$$\ln [X_{Cu}]_{sat} = -\frac{8060.5}{T} + 5.207 \qquad \text{(E.11.14)}$$

Calculate the maximum extent of removal of copper by this reaction at 800°C.

Solution   For the reaction (E.11.13),

$$\Delta G_{13}^0 = - RT \ln K_{13} = - RT \ln \left\{ \frac{(a_{Cu_2S}) \times [a_{Pb}]}{[a_{Cu}]^2 \times (a_{PbS})} \right\}_{eq} \tag{E.11.15}$$

Since Cu_2S and PbS are pure, $a_{Cu_2S} = a_{PbS} = 1$. Concentration of Cu in liquid Pb would be very small (can be verified later). Hence Pb is almost pure, and the activity of Pb may be assumed as 1. Therefore,

$$\Delta G_{13}^0 = - RT \ln \left\{ \frac{1}{[a_{Cu}]^2} \right\}_{eq} \tag{E.11.16}$$

Assuming that Henry's Law is applicable to Cu in liquid Pb, we have

$$a_{Cu} = \gamma_{Cu}^0 \cdot X_{Cu} = \frac{X_{Cu}}{[X_{Cu}]_{sat}} \tag{E.11.17}$$

on the basis of (Eq. 10.59).
From Eq. (E.11.14),

$$[X_{Cu}]_{sat} = 0.10 \text{ at } 800°C \text{ (i.e. 1073 K)}$$

$$\Delta G_{13}^0 = \left[\Delta G_f^0 (Cu_2S) - \Delta G_f^0 (PbS) \right]$$

$$= [(-142,900 - 11.3T \ln T + 120.2T) - (-157,250 + 80T)]$$

$$= 14,350 - 11.3T \ln T + 40.2T = - 27,126 \text{ J mol}^{-1} \text{ at 1073 K.}$$

Combining Eqs. (E.11.16) and (E.11.17) and putting in values, $X_{Cu} = 0.0216$.

This is the minimum copper concentration attainable. In practice, it may be somewhat higher if equilibrium is not attained.

Since $X_{Pb} = 1 - X_{Cu} = 0.9784 \simeq 1$, the assumption that $a_{Pb} = 1$ is all right here without much error. If it were not the case, then the calculation had to be repeated once more by assuming Raoult's Law for Pb (i.e. $a_{Pb} = 1 - X_{Cu}$).

11.4   Interactions amongst Solutes in Dilute Multicomponent Solutions

In a binary A-B solution, the interactions amongst components may be classified into the following three categories: A-B, A-A, B-B. In a ternary A-B-C, the interaction pairs would be: A-B, A-C, B-C, A-A, B-B, C-C. These additional interactions will have some effect on activity coefficients. Now take a binary solution. Suppose the activity coefficient of B is γ_B; add a component C to the solution; the value of γ_B will change to some other value (say γ_B') due to additional interaction of component C on γ_B.

In metallurgy and materials science, we have to often deal with multicomponent solutions—an important issue. In concentrated multicomponent solutions, the mathematical procedure for dealing with such interactions is lengthy, and is beyond the scope of an introductory thermodynamics course. For example, during extraction and refining of liquid metals, we mostly deal with dilute multicomponent solutions. An important example is liquid steel, where Fe is the solvent, and several solutes (C, Si, Mn, S, P etc.) are present in very low concentrations. For such dilute multicomponent solutions, Carl Wagner (1952) proposed a simple but elegant analytical method of handling solute-solute interactions, which is discussed now.

11.4.1 Interaction Coefficients

Let A designate the solvent, and 1, 2, ... i, j, ... are the symbols for the solutes. Wagner expressed $\ln \gamma_i$ (i = general symbol of solute) as function of X_1, X_2, ... by the Taylor series expansion around pure solvent A. The equation is as follows:

$$\ln \gamma_i \text{ (multicomponent)} = \ln \gamma_i^0 + \left[X_1 \left(\frac{\delta[\ln \gamma_i]}{\delta X_1} \right)_{X_A \to 1} + ... + X_j \left(\frac{\delta[\ln \gamma_i]}{\delta X_j} \right)_{X_A \to 1} + ... \right]$$

$$+ \left[\frac{1}{2} X_1^2 \left(\frac{\delta^2[\ln \gamma_i]}{\delta X_1^2} \right)_{X_A \to 1} + ... + \frac{1}{2} X_j^2 \left(\frac{\delta^2[\ln \gamma_i]}{\delta X_j^2} \right)_{X_A \to 1} + ... \right] \quad (11.18)$$

Here, γ_i^0 = Henry's Law constant in $A - i$ binary.

For dilute solution, X_1, X_2, ... X_j ... are very small. Hence, Wagner ignored all higher order terms. Then Eq. (11.18) becomes

$$\ln \gamma_i = \ln \gamma_i^0 + X_1 \left(\frac{\delta[\ln \gamma_i]}{\delta X_1} \right)_{X_A \to 1} + ... + X_i \left(\frac{\delta[\ln \gamma_i]}{\delta X_i} \right)_{X_A \to 1} + X_j \left(\frac{\delta[\ln \gamma_i]}{\delta X_j} \right)_{X_A \to 1} + ...$$

$$(11.19)$$

Although $X_A \to 1$ is the mathematically rigorous constraint, it is valid only if mole fractions of all solutes tend to zero. On the basis of this constraint, Eq. (11.19) may be rewritten as

$$\ln \gamma_i = \ln \gamma_i^0 + X_1 \left(\frac{\delta[\ln \gamma_i]}{\delta X_1} \right)_{X_1 \to 0} + ... + X_i \left(\frac{\delta[\ln \gamma_i]}{\delta X_i} \right)_{X_i \to 0} + X_j \left(\frac{\delta[\ln \gamma_i]}{\delta X_j} \right)_{X_j \to 0} + ...$$

$$(11.20)$$

The influence of one solute on γ of another solute was termed as *interaction coefficient* by Wagner. This allows us to rewrite of Eq. (11.20) as

$$\ln \gamma_i = \ln \gamma_i^0 + X_1 \varepsilon_i^1 + X_2 \varepsilon_i^2 + \ldots + X_i \varepsilon_i^i + X_j \varepsilon_i^j + \ldots = \ln \gamma_i^0 + \sum_j X_j \varepsilon_i^j \qquad (11.21)$$

where

$$\varepsilon_i^j = \left(\frac{\delta [\ln \gamma_i]}{\delta X_j} \right)_{X_j \to 0} \qquad (11.22)$$

and ε_i^j is known as *interaction coefficient describing influence of solute j on* $\ln \gamma_i$.

11.4.2 Equality of ε_i^j and ε_j^i

In analogy with the definition of ε_i^j in Eq. (11.22),

$$\varepsilon_j^i = \left(\frac{\delta \left[\ln \gamma_j \right]}{\delta X_i} \right)_{X_i \to 0} \qquad (11.23)$$

where ε_j^i = interaction coefficient describing the influence of solute i on $\ln \gamma_j$.

From Eq. (9.11) we may write, at constant P and T,

$$\bar{G}_i = \left(\frac{\delta G'}{\delta n_i} \right)_{m_1 \ldots, \text{except } n_i} , \qquad \bar{G}_j = \left(\frac{\delta G'}{\delta n_j} \right)_{m_1 \ldots, \text{except } n_j} \qquad (11.24)$$

Again,

$$\bar{G}_i^m = \left(\frac{\delta (G' - G_0')}{\delta n_i} \right)_{m_1 \ldots, \text{except } n_i} , \qquad \bar{G}_j^m = \left(\frac{\delta (G' - G_0')}{\delta n_j} \right)_{m_1 \ldots, \text{except } n_j} \qquad (11.25)$$

where G_0' is G' when components 1, 2, …, i, j, … are present as pure in mixture of the same overall composition.

From Eq. (11.25),

$$\frac{\delta \bar{G}_i^m}{\delta n_j} = \frac{\delta^2 (G' - G_0')}{\delta n_i \, \delta n_j} = \frac{\delta \bar{G}_j^m}{\delta n_i} \qquad (11.26)$$

Again,

$$\left(\frac{\delta \bar{G}_i^m}{\delta n_j} \right)_{X_j \to 0} = RT \left[\frac{\delta (\ln a_i)}{\delta n_j} \right]_{X_j \to 0} = \frac{RT}{n_T} \left\{ \frac{\delta (\ln X_i) + \delta (\ln \gamma_i)}{\delta X_j} \right\}_{X_j \to 0}$$

$$= \frac{RT}{n_T} \left[\frac{\delta (\ln \gamma_i)}{\delta X_j} \right]_{X_j \to 0} = \frac{RT}{n_T} \varepsilon_i^j \qquad (11.27)$$

since X_i and X_j are independent variables. Similarly,

$$\left(\frac{\delta \overline{G}_j^m}{\delta n_i} \right)_{X_i \to 0} = \frac{RT}{n_T} \, \varepsilon_j^i \tag{11.28}$$

Combining Eqs. (11.26)–(11.28), we get

$$\varepsilon_i^j = \varepsilon_j^i \tag{11.29}$$

This is a very important conclusion. It means that if experimental value of ε_i^j is available, we automatically know ε_j^i.

11.4.3 Practical Utility of Interaction Coefficients

As already stated, activity versus composition data are to be basically obtained from experimental measurements. For a binary solution, it is the simplest. The activity of one component is measured, and from G–D integration, the activity of the other can be calculated. This approach works for ternary also. For multicomponent solutions, however, much larger experimental programme is required since there are several composition variables. This is where the concept of Interaction coefficient for dilute solutions comes in handy.

Let us consider the multicomponent A, 1, 2, ... $i, j,$ ... again. For knowing γ_i, what we have to do now is to make measurements in binary A-i, and all ternaries, i.e. A-i-1, A-i-2, ... A-i-j, ... Experimental data in A-i allows determination of γ_i^0. If experimental data in A-i and A-i-j, are combined, we can find out ε_i^j since in the ternary A-i-j,

$$\ln \gamma_i = \ln \gamma_i^0 + X_j \varepsilon_i^j \tag{11.30}$$

Similarly, from data on other ternaries, $\varepsilon_i^1, \varepsilon_i^2$ ... are evaluated. Then these can be employed to calculate γ_i in multicomponent solution by Eq. (11.21). This is a tremendous saving of experimental efforts.

ε_i^i is known as *self-interaction coefficient and is a measure of deviation from Henry's Law in A-i binary. If HL is obeyed, then* $\varepsilon_i^i = 0$.

Although derivation of Eq. (11.21) assumes mole fractions of solutes as tending to zero, it has been found that often the equation can be satisfactorily employed upto a reasonable concentration. For still higher concentration, retention of the *second order terms* is necessary. For example,

$$r_i^j = \left[\frac{\delta^2 (\ln \gamma_i)}{\delta X_j^2} \right]_{X_j \to 0} \tag{11.31}$$

is the *second order interaction coefficient* describing influence of j on $\ln \gamma_i$.

11.5 One Weight Per cent Standard State

11.5.1 Introduction

The conventional standard state is also known as Raoultian standard state, where the standard state (SS) is pure solid or pure liquid, whichever is the stablest at the temperature under consideration. Moreover, *Raoult's Law serves as the reference.*

Thermodynamics permits use of any other alternative standard state. Such a state may *be real or hypothetical.* However, for such a change of SS, the thermodynamic data employed for calculation should conform to the new SS. For example, ΔG^0 for reactions in general data sources are all for the conventional SS. They are to be changed if a new SS is employed.

In extraction and refining of liquid metals as well as for a variety of other uses, we deal with dilute solutions. Also, industries prefer to find out answers directly in weight percent. In order to take care of these, Chipman proposed the "1 wt.% Standard State". Its features are:

- Composition scale is in wt.%, not in mole fraction.
- Henry's Law, rather than Raoult's Law, is the basis.
- The answer for equilibrium calculation would be directly in wt.%.

Consider the multicomponent solution with 1, 2, ..., i, j, ... as components. Then conversion of wt.% into mole fraction can be done with the help of the following equation:

$$X_i = \frac{W_i/M_i}{\Sigma(W_i/M_i)}$$

(11.32)

where

W_i = wt.% of i,

M_i = molecular mass of i.

The reverse conversion is as follows:

$$W_i = \frac{100\,X_i M_i}{\Sigma X_i M_i}$$

(11.33)

11.5.2 Definition and Relations for 1 wt.% Standard State

Consider a dilute binary solution A-B, where A is the solvent and B the solute. Then the activity of B in 1 wt.% standard state is defined as

$$h_B = \frac{a_B}{\left(\dfrac{a_B}{W_B}\right)_{W_B \to 0}}$$

(11.34)

where h_B is activity of B in the new SS.

When $W_B \to 0$, Henry's Law (HL) is obeyed. Hence, $\left(\dfrac{a_B}{W_B}\right)_{W_B \to 0}$ is constant and is

related to HL constant (γ_B^0). Now, if we consider the application of the above equation on the HL line, then a_B in both the numerator and the denominator in Eq. (11.34) are the same, and Eq. (11.34) gets simplified into

$$h_B = W_B \text{ at } W_B \to 0 \tag{11.35}$$

In other words, on HL line, $h_B = W_B$.

Now, consider the point on the HL line at $W_B = 1$. From Eq. (11.35), $h_B = 1$ on that point. *This is the 1 wt.% standard state.* The SS may be real or hypothetical. Figure 11.5(a) and 11.5(b) illustrate these.

In real SS, HL extends beyond $W_B = 1$ (Fig. 11.5a). In the hypothetical SS, HL ends below $W_B = 1$ (Fig. 11.5b). At point S, $h_B = 1$, and hence it corresponds to 1 wt.% standard state.

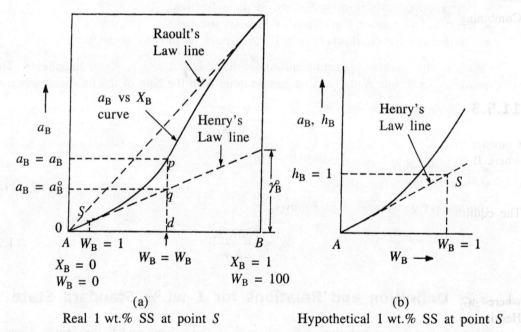

(a)   (b)

Real 1 wt.% SS at point S Hypothetical 1 wt.% SS at point S

Fig. 11.5 Illustration for 1 wt.% SS: (a) real 1 wt.% SS, and (b) hypothetical 1 wt.% SS.

In Fig. 11.5(a), at $W_B = W_B$, $pd = h_B$, $qd = W_B$ for 1 wt% SS. Hence,

$$\frac{pd}{qd} = \frac{h_B}{W_B} = f_B \tag{11.36}$$

where $f_B = $ *activity coefficient of B in 1 wt.% SS.*

From the value of γ_B^0 shown in Fig. 11.5(a), at point q,

$$a_B = a_B^0 = \gamma_B^0 X_B \tag{11.37}$$

Since at p, $a_B = \gamma_B X_B$, the combination of Eqs. (11.36) and (11.37) yields

$$f_B = \frac{h_B}{W_B} = \frac{pd}{qd} = \frac{a_B}{a_B^0} = \frac{\gamma_B X_B}{\gamma_B^0 X_B} = \frac{\gamma_B}{\gamma_B^0} \tag{11.38}$$

Again, from Eq. (11.38),

$$\frac{h_B}{a_B} = \frac{W_B}{a_B^0} = \frac{W_B}{\gamma_B^0 X_B} \tag{11.39}$$

For binary A-B, and $W_B \to 0$, $W_A \to 100$, Eq. (11.32) gets simplified into

$$X_B \simeq \frac{W_B M_A}{100 M_B} \tag{11.40}$$

Combining the above, we get

$$\frac{h_B}{a_B} \simeq \frac{100 M_B}{\gamma_B^0 M_A} \tag{11.41}$$

11.5.3 Evaluation of ΔG^0 for Reaction Involving 1 wt.% SS

Consider the following reaction at temperature T for a *dilute binary metallic solution* A-B, where B is the solute.

$$[B]_{\text{dil. A-Bsoln}} + O_2(g) = (BO_2) \tag{11.42}$$

The equilibrium relation is to be rewritten, for 1 wt.% SS for B, as

$$\Delta G_h^0 = - RT \ln K_h = - RT \ln \left[\frac{(a_{BO_2})}{[h_B] \, p_{O_2}} \right]_{eq} \tag{11.43}$$

where ΔG_h^0, K_h are ΔG^0 and K for 1 wt.% SS for component B. The subscript 'h' denotes Henrian since HL is reference here. The corresponding equation in the Raoultian SS are:

$$\Delta G^0 = - RT \ln K = - RT \ln \left\{ \frac{(a_{BO_2})}{[a_B] \, p_{O_2}} \right\}_{eq} \tag{11.44}$$

From Eqs. (11.43) and (11.44),

$$\Delta G_h^0 - \Delta G^0 = - RT \ln \left(\frac{K_h}{K} \right) = - RT \ln \left(\frac{a_B}{h_B} \right) = RT \ln \left(\frac{h_B}{a_B} \right) \tag{11.45}$$

Combining Eq. (11.45) with Eq. (11.41), we get

$$\Delta G_h^0 = \Delta G^0 + RT \ln \left(\frac{100\, M_B}{\gamma_B^0\, M_A} \right) \tag{11.46}$$

Equation (11.46) provides the basis for evaluation of ΔG_h^0. ΔG^0 is available in standard thermochemical data sources. Hence, experimental determination of γ_B^0 in binary A-B is the only additional information required. γ_B^0 depends on both B and A. Such experimental data are available with important solvents such as liquid Fe, Cu, Pb, and a variety of solutes. Tables are also available with compilation of ΔG_h^0 for the formation of common compounds.

11.5.4 Dilute Multicomponent Solution in 1 wt.% SS

In multicomponent solutions, additional solute-solute interactions are to be considered. It has been done elaborately for Raoultian standard state in Section 11.4. For 1 wt.% SS, Eq. (11.21) has been modified as

$$\log f_i = \log f_i^0 + \sum_j W_j e_j^i \tag{11.47}$$

f_i^0 is Henry's Law constant for 1 wt. % SS. Since $f_i^0 = 1$ by definition, $\log f_i^0 = 0$. Hence, Eq. (11.47) gets simplified into

$$\log f_i = \sum_j W_j e_i^j = W_1 e_i^1 + W_2 e_i^2 + \ldots + W_i e_i^i + W_j e_i^j + \ldots \tag{11.48}$$

The common logarithm was chosen for convenience, and

$$e_i^j = \left[\frac{\delta(\log f_i)}{\delta W_j} \right]_{W_j \to 0} \tag{11.49}$$

The relationship of e_i^j with e_j^i may be derived on the basis of Section 11.4, as follows:
Combining Eq. (11.48) with Eq. (11.38), the former may be rewritten as

$$\ln f_i = \ln \left(\frac{\gamma_i}{\gamma_i^0} \right) = 2.303 \sum_j W_j e_i^j \tag{11.50}$$

i.e.

$$\ln \gamma_i = \ln \gamma_i^0 + 2.303 \sum_j W_j e_i^j \tag{11.51}$$

A term-by-term comparison with Eq. (11.21) leads to

$$2.303 W_j e_i^j = X_j \varepsilon_i^j \tag{11.52}$$

Combining Eq. (11.52) with (11.40), we obtain

$$e_i^j = \frac{X_j}{2.303 \, W_j} \, \varepsilon_i^j = \frac{W_j M_A}{100 \, M_j \, 2.303 \, W_j} \, \varepsilon_i^j = \frac{M_A}{230.3 \, M_j} \, \varepsilon_i^j \tag{11.53}$$

This allows evaluation of e_i^j from ε_i^j and vice versa. Similarly,

$$e_j^i = \frac{M_A}{230.3 \, M_i} \, \varepsilon_j^i \tag{11.54}$$

Since $\varepsilon_i^j = \varepsilon_j^i$, from Eq. (8.104), we have

$$e_j^i = \frac{M_A}{230.3 \, M_i} \, \varepsilon_j^i = \frac{M_A}{230.3 \, M_i} \, \frac{230.3 \, e_i^j \, M_j}{M_A} = \frac{M_j}{M_i} \, e_i^j \tag{11.55}$$

For a ternary A-i-j, Eq. (11.48) reduces to: $\log f_i = W_j e_i^j$. Hence, plots of $\log f_i$ vs. W_j would be straight lines passing through the origin. Figure 11.6 shows such plots for activity of nitrogen dissolved in liquid iron at 1833 K for several ternaries such as Fe-N-Ni and Fe-N-Cr. Nitrides of Cr, V etc. are stronger than nitride of Fe. In other words, Cr-N, V-N bonds are stronger than Fe-N bonds, whereas Ni-N bond is weaker than Fe-N bond. This explains why f_N decreases upon addition of Cr etc., and increases upon addition of elements like Ni.

Tables of ε_i^j, e_i^j in metallic solutions for common solvents (Fe, Cu etc.) at high temperature are available in literature.

EXAMPLE 11.5 A H_2–H_2S gas mixture with H_2/H_2S ratio of 10 was equilibrated with pure liquid iron at 1850 K. On analysis of quenched sample, the iron was found to contain 0.8 wt.% sulphur. Calculate the free energy change for the following reaction at 1850 K:

$$\frac{1}{2} S_2 \text{ (g, 1 atm)} = [S]_{1 \text{ wt.\% SS in liq. Fe}} \tag{E.11.18}$$

Given at 1850 K: $\log f_S = -0.028 W_S$

$$H_2(g) + \frac{1}{2} S_2(g) = H_2S(g), \quad \Delta G_{19}^0 = 690 \text{ J mol}^{-1} \tag{E.11.19}$$

Solution

$$H_2(g) + [S]_{1 \text{ wt\%SS}} = H_2S(g); \quad \Delta G_{20}^0 \tag{E.11.20}$$

Substracing Eq. (E.11.20) from (E.11.19), we get

$$\frac{1}{2} S_2(g) = [S]_{1 \text{ wt\% SS}}, \quad \Delta G_{21}^0 = \Delta G_{19}^0 - \Delta G_{20}^0 \tag{E.11.21}$$

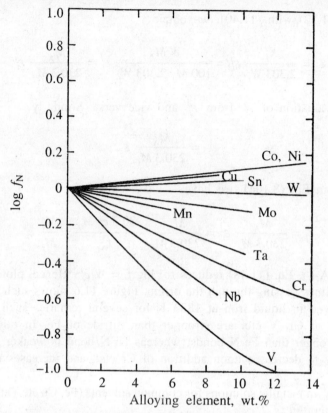

Fig.11.6 Activity coefficients of nitrogen in Fe-N-j ternary liquid solutions at 1600°C, with iron as the solvent [R.G. Ward, *Physical Chemistry of Iron and Steelmaking*, Edward Arnold, London, 1962].

Now,

$$\Delta G_{20}^0 = - RT \ln \left\{ \frac{p_{H_2S}}{p_{H_2} \times [h_S]} \right\}_{eq} = - RT \ln \left(\frac{1}{10 \, [h_S]} \right)_{eq} \qquad (E.11.22)$$

Again,

$$h_S = f_S W_S = (10^{-0.028 \times 0.8}) \cdot 0.8 = 0.76$$

Putting the above value of h_S in Eq. (E.11.22), we obtain

$$\Delta G_{20}^0 = 31,190 \text{ J mol}^{-1}, \qquad \Delta G_{21}^0 = 690 - 31,190 = - 30,500 \text{ J mol}^{-1}$$

EXAMPLE 11.6   From experimental measurements, it has been determined that, for Si (pure, l) = [Si]$_{1 \text{ wt\% SS in liq. Fe}}$,

$$\bar{G}_{Si}^m = - 119.24 \times 10^3 - 24T \text{ J mol}^{-1} \qquad (E.11.23)$$

At 1600°C (i.e. 1873 K), γ_{Si} relative to pure Si as standard state is 0.0014 at $X_{Si} = 0.01$.

Calculate f_{Si} at $X_{Si} = 0.01$. Also calculate $\bar{H}_{Si}^m$ at 1 wt.% standard state of Si in liquid Fe, at infinite dilution of Si.

(i) $$\bar{G}_{Si}^m = RT \ln [a_{Si}]_{at\ 1\ wt\ \%\ SS} = RT \ln (\gamma_{Si}^0 [X_{Si}]_{at\ 1\ wt\ \%}) \qquad (E.11.24)$$

since 1 wt.% SS is, by definition, on Henry's Law line. Again, from Eq. (11.40), at $W_{Si} = 1$,

$$X_{Si} = \frac{M_{Fe}}{M_{Si} \times 100} = \frac{56}{2800} = 0.02.$$

Solving Eq. (E.11.24), we get $\gamma_{Si}^0 = 1.25 \times 10^{-3}$. At $X_{Si} = 0.01$, $\gamma_{Si} = 0.0014$; from Eq. (11.50),

$$f_{Si} = \frac{\gamma_{Si}}{\gamma_{Si}^0} = \frac{0.0014}{1.25 \times 10^{-3}} = 1.12$$

[*Note:* Here, Henry's Law is not obeyed at $X_{Si} = 0.01$ since $f_{Si} \neq 1$. Hence, obviously HL would not be obeyed at 1 wt.% (i.e. $X_{Si} = 0.02$). This means that the 1 wt.% SS is not real but hypothetical.]

(ii) $$\bar{H}_{Si}^m = \left[\frac{\delta \left(\dfrac{\bar{G}_{Si}^m}{T} \right)}{\delta \left(\dfrac{1}{T} \right)} \right]_{P,\ comp} \qquad (E.11.25)$$

according to Gibbs–Helmholtz equation.

At infinite dilution, HL is obeyed and $\bar{G}_{Si}^m$ is given by the Eq. (E.11.23). Hence,

$$\bar{H}_{Si}^m = \frac{\delta \left(-\dfrac{119.24 \times 10^3 + 24.0T}{T} \right)}{\delta \left(\dfrac{1}{T} \right)} = -119.24 \times 10^3 \text{ J mol}^{-1}$$

EXAMPLE 11.7 Liquid iron containing 2 wt.% silicon, 2 wt.% chromium, and 0.5 wt.% vanadium was kept in a graphite crucible at 1600°C for a long time till equilibrium was established between the liquid and the crucible. Calculate the wt.% of carbon in iron after the experiment is over.

Given:

(i) Solubility of graphite in pure liquid iron = 5.4 wt.% at 1600°C

(ii) For Fe-C binary, $\log f_C^C = \dfrac{4350}{T} [1 + 4 \times 10^{-4}(T - 1770)] (1 - X_{Fe})^2$ \qquad (E.11.26)

(iii) $e_C^{Si} = 0.1$, $e_C^{Cr} = -0.024$, $e_C^V = -0.033$

Solution $\qquad\qquad\qquad T = 1600 + 273 = 1873 \text{ K}$

At graphite saturation, $[a_C] = 1$ in liquid Fe, i.e., $X_C \gamma_C = X_C \gamma_C^0 f_C = 1$ $\qquad$ (E.11.27)

(i) In binary Fe-C solution,

$$X_C = \frac{W_C/12}{\dfrac{W_C}{12} + \dfrac{W_{Fe}}{56}} = \frac{5.4/12}{\dfrac{5.4}{12} + \dfrac{100 - 5.4}{56}} = 0.21$$

Noting that

$$1 - X_{Fe} = X_C, \ \log f_C^C = \frac{4350}{1873}[1 + 4 \times 10^{-4} (1873 - 1770)] (0.21)^2 = 0.1067$$

From Eq. (E.11.27), we have, $\gamma_C^0 = 3.716$.

(i) For liquid Fe-Cr-V-Si-C solution,

$$\log f_C = \log f_C^C + e_C^{Si} W_{Si} + e_C^{Cr} W_{Cr} + e_C^{V} W_{V}$$

$$= \log f_C^C + 0.1 \times 2 - 0.024 \times 2 - 0.033 \times 0.5$$

$$= \log f_C^C + 0.1355 \qquad\qquad\qquad (E.11.28)$$

[*Note:* f_C^C in Fe-C binary and in the multicomponent solution are different since the value of X_{Fe} will be different. Since the interaction effects of Si, Cr, V have been separated, from a thermodynamic point of view, approximately speaking, $(1 - X_{Fe})$ can be substituted by X_C in Eq. (E.11.26)]. Hence, from Eq.(E.11.26),

$$\log f_C^C = 2.322[1 + 412 \times 10^{-4}] X_C^2 = 2.418 X_C^2 \qquad (E.11.29)$$

Combining all the above equations, we get

$$3.716 \, X_C 10^{(2.418 X_C^2 + 0.1355)} = 1 \qquad\qquad (E.11.30)$$

Graphical solution yields $X_C = 0.169$. Again,

$$X_C = \frac{W_C/12}{\dfrac{W_{Si}}{28} + \dfrac{W_{Cr}}{52} + \dfrac{W_C}{12} + \dfrac{W_V}{51} + \dfrac{W_{Fe}}{56}} \qquad (E.11.31)$$

Since $W_{Si} = 2$, $W_{Cr} = 2$ and $W_V = 0.5$, we have

$$W_{Fe} = 96.5 - W_C \qquad\qquad\qquad (E.11.32)$$

Combining the above equations, and further graphical solution yields $W_C = 4.2$ wt.% after crucible-melt equilibrium is attained.

11.6 Sieverts' Law

Sieverts carried out extensive measurements of solubilities of diatomic gases (H_2, N_2 etc.) in metals. Based on these, he proposed that solubilities of these gases are proportional to square roots of partial pressures at a constant temperature. This is readily explained if we recognize that all solutes are present in atomic state (not as molecules) in a metallic solution. Therefore, dissolution of nitrogen follows the reaction

$$\frac{1}{2} N_2(g) = [N]_{\text{in metal}} \tag{11.56}$$

for which

$$K_h = \left(\frac{[h_N]}{p_{N_2}^{1/2}} \right)_{eq} \tag{11.57}$$

Since solubility means concentration of N in metallic solution at equilibrium with a certain value of p_{N_2}, Eq. (11.57) is applicable. For binary Metal-N solutions, solubilities of nitrogen is very small. Hence Henry's Law is applicable, and on the basis of Eq. (11.35), $h_N = W_N$. Therefore,

$$K_h = \left(\frac{[W_N]}{p_{N_2}^{1/2}} \right)_{eq} , \text{ i.e. } [W_N]_{\text{at solubility limit}} = K_h p_{N_2}^{1/2} \tag{11.58}$$

11.7 Summary

1. This chapter contains discussions of some additional topics to supplement the earlier chapters so that the readers have complete information for solution of all types of equilibrium calculations.

2. As an extension of Chapter 8, a further analysis of phase stabilities of reactive stoichiometric compounds has been presented with the example of construction of Si-C-O ternary diagram. Wustite (Fe_xO) is the best metallurgical example of a nonstoichiometric compound. Thermodynamic analysis of this has been discussed in the chapter.

3. Reaction equilibria calculations involving solutions of condensed phases have been illustrated by solved examples.

4. In a multicomponent solution, components interact amongst themselves. For a dilute solution, in connection with interactions amongst solutes, Wagner derived analytical equation to quantify such interactions, and the equation is

$$\ln \gamma_i = \ln \gamma_i^0 + \sum_j X_j \varepsilon_i^j$$

where γ_i is activity coefficient of i in the multicomponent solution with solvent A and solutes, 1, 2, ..., i, j, ... γ_i^0 is Henry's Law constant in binary A-i solution. ε_i^j is the interaction coefficient describing the influence of j on γ_i, and is given as

$$\varepsilon_i^j = \left(\frac{\delta[\ln \gamma_i]}{\delta X_j} \right)_{X_j \to 0}$$

Moreover, $\varepsilon_j^i = \varepsilon_i^j$.

5. Weight% is commonly used as the composition parameter for practical applications. In order to obtain answers of calculations directly in wt.%, the 'one weight percent standard state' was proposed. It is based on Henry's Law (HL) as reference. On HL line, in the binary A-B, where B is the solute,

Activity of B in 1 wt.% SS (h_B) = W_B, where W_B is wt.% B.

At a composition where HL is not obeyed,

$$h_B = f_B W_B$$

where f_B = activity coefficient of B in 1 wt.% SS.

6. Use of 1 wt.% SS requires employment of a revised value of free energy of reaction (ΔG_h^0), which can be calculated from experimental values of ΔG^0 and γ_B^0 (i.e. HL constant in binary A-B). Thermodynamic data tables are available for ΔG_h^0 most extensively with liquid iron as solvent.

7. For multicomponent dilute solution, in 1 wt.% SS,

$$\log f_i = \log f_i^0 + \sum_j W_j e_i^j$$

where

$$e_i^j = \left(\frac{\delta[\log f_i]}{\delta W_j} \right)_{W_j \to 0} , \text{ and } \log f_i^0 = 0 \text{ by definition.}$$

PROBLEMS

11.1 A methane-hydrogen gas mixture at 1 atm. total pressure is allowed to come to equilibrium with a steel containing 0.60 wt.% carbon at 925°C: The gas was found to contain 0.64% CH_4 and the rest H_2 by volume. Calculate

   (i) the activity of carbon in steel;

   (ii) the expected analysis of a CO–CO_2 gas mixture which would be at equilibrium with carbon in steel.

11.2 An alloy with 20 atom% Ni and 80 atom%. Au is allowed to react with a gas mixture of H_2O-H_2 at 1000 K to form pure solid NiO. At equilibrium the gas contains 0.175% H_2, 50% N_2, and the rest H_2O. Find the activity of Ni in the alloy.

11.3 A Cr-M solid solution contains 15 atom% Cr, and behaves ideally. Upon reaction with a H_2O-H_2 gas mixture, only pure solid Cr_2O_3 forms. Calculate the p_{H_2}/p_{H_2O} ratio in the gas mixture at equilibrium with the solid solution and Cr_2O_3 at 1000°C.

11.4 Consider the reaction between a liquid sulphide solution of Cu_2S-FeS (known as Matte) and a molten slag of FeO-SiO_2-Cu_2O at 1250°C. Calculate the activity of Cu_2O in the slag upon attainment of equilibrium. The matte contains 40 wt.% Cu_2S and the rest FeS, and behaves as an ideal solution. The activity of FeO in the slag may be assumed as 0.5 relative to pure liquid FeO.

11.5 Zinc, present as impurity in molten lead, can be selectively removed as $ZnCl_2$ by treatment with chlorine gas. Find the residual concentration of Zn in liquid Pb upon attainment of the following reaction equilibrium at 663 K:

$$[Zn] + (PbCl_2) = (ZnCl_2) + Pb(l)$$

Given: (i) $ZnCl_2$-$PbCl_2$ forms an ideal liquid solution.

      (ii) ΔG^o for formation of $PbCl_2$ and $ZnCl_2$ are -257.5 kJ mol^{-1} and -320.7 kJ mol^{-1}, respectively at 663 K.

      (iii) X_{ZnCl_2} in salt phase = 0.983.

      (iv) $\ln [\gamma^0_{Zn}]_{in\ Pb} = \dfrac{2798}{T} - 0.962$

11.6 It is also possible to lower Zn content in liquid Pb by vacuum distillation. Calculate the maximum pressure that can be tolerated in the vacuum system at 900 K for the following conditions:

    (i) Zn as only solute in liquid Pb, and Zn concentration is to be lowered to 0.01 atom%.

    (ii) Liquid Pb contains 2 wt.% cadmium and 2 wt.% copper, besides Zn. The Zn content is to be lowered to 0.3 wt.%.

Given: (i) $e^{Cd}_{Zn} = -0.04$, $e^{Cu}_{Zn} = -0.07$, $e^{Zn}_{Zn} = 0$ in liquid Pb

      (ii) γ^0_{Zn} in liquid Pb as in Problem 11.5.

11.7 An Fe-Mn solid solution containing $X_{Mn} = 0.001$ is in equilibrium with an FeO-MnO solid solution and an oxygen containing gaseous atmosphere at 1000 K. Calculate: (i) Composition of the oxide solid solution, and (ii) p_{O_2} in the gaseous atmosphere. Both the metallic and oxide solution are ideal.

11.8 A mixture of solid CaO, MgO, $3CaO.Al_2O_3$ and liquid aluminium exerts an equilibrium vapour pressure of magnesium of 0.035 atm. at 1300 K. Write the equation for the appropriate reaction equilibrium. Calculate the standard free energy of formation of $3CaO.Al_2O_3$ from CaO and Al_2O_3, and the activity of Al_2O_3 in CaO-saturated $3CaO.Al_2O_3$ at 1300 K.

11.9 A liquid iron solution contains 0.5 wt.% silicon and 2 wt.% carbon. Temperature is 1900 K. Calculate (i) h_{Si}, (ii) μ_{Si}. Given: γ^0_{Si} in binary Fe-Si solution $= 1.37 \times 10^{-3}$;

$$e^C_{Si} = 0.18, \qquad e^{Si}_{Si} = 0.11$$

11.10 (i) In the binary Fe-Cu liquid solution at 1550°C with Fe as the solvent, it is given that:

$$\log \gamma_{Cu} = 1.45 \, X^2_{Fe} - 1.86 \, X^3_{Fe} + 1.41 \, X^4_{Fe}$$

Calculate the free energy change accompanying the change of standard state of Cu from pure liquid Cu to 1 wt.% SS in liquid Fe.

(ii) What would be the corresponding free energy change for Mn in Fe-Mn liquid solution at 1550°C, which behaves as an ideal solution?

11.11 An iron –1 wt.% vanadium liquid alloy is equilibrated at 1900 K with a steam-hydrogen mixture containing 5% steam. On analysis, the metal was found to contain 0.033 wt.% oxygen. Assuming that oxygen dissolved in molten iron obeys Henry's Law, calculate e^V_0.

11.12 An iron-titanium liquid solution with Fe as the solvent is at equilibrium with solid pure TiO_2 and a H_2O-H_2 gas mixture having p_{H_2O}/p_{H_2} ratio of 5.56×10^{-3}. Temperature is 1600°C, and $h_{Ti} = 1$. Calculate γ^0_{Ti}.

11.13 Calculate the residual oxygen content of liquid iron containing 0.1 wt.% Si, at equilibrium with pure solid silica at 1600°C. $e^{Si}_{Si} = +0.32$, $e^0_{Si} = -0.24$, $e^0_0 = -0.20$ (since e^{Si}_0 can be calculated, it is not given).

11.14 When an iron-phosphorus liquid solution is equilibrated at 1900 K with solid CaO, solid 3CaO. P_2O_5, and a gas phase containing $p_{O_2} = 10^{-10}$ atm, the activity of P in the iron with respect to 1 wt.% ss in Fe is 20. Calculate ΔG^0_{1900} for the reaction:

$$P_2(g) + \frac{5}{2} O_2(g) = P_2O_5(g)$$

Given: (i) $3CaO(s) + P_2O_5(g) = 3CaO.P_2O_5(s)$

$$\Delta G^0_{1900} = -564,000 \ J \ mol^{-1}$$

(ii) $1/2 \ P_2(g) = [P]_{1 \ wt\% \ SS \ in \ Fe}$

$$\Delta G^0 = -122,200 - 19.2 \ T \ J \ mol^{-1}$$

11.15 Calculate the chemical potential of nitrogen gas (i.e. N_2) in liquid steel at 1600°C. The steel has the following composition:

$$W_N = 0.01, \qquad W_C = 0.5, \qquad W_P = 0.2, \qquad W_{Mn} = 0.5$$

Given: (i) $[N]_{1 \ wt\% \ SS} = K_N p^{1/2}_{N_2}, \quad \log K_N = -\dfrac{188.1}{T} - 1.246$

(ii) $e^C_N = 0.25, \quad e^N_N = 0, \quad e^P_N = 0.051, \quad e^{Mn}_N = -0.02$

Chapter *12*

Third Law of Thermodynamics, Statistical Thermodynamics, and Entropy

As discussed in Section 4.4, although Clausius proposed entropy in 1850, it took several more decades to fully establish the fundamental significance of entropy qualitatively and quantitatively from the atomistic viewpoint. Early work leading to the Third Law of Thermodynamics was done by T.W. Richards (1902). W. Nernst (1906) generalized these findings and formulated the statement of the Third Law, which he called as *Heat Theorem*. Subsequently, the statement was refined by Max Planck. The Third Law provided some atomistic interpretation of entropy qualitatively. It specifically dealt with the behaviour of entropies and specific heats of substances at absolute zero of temperature (i.e. $T = 0$).

The subject of Statistical Thermodynamics started developing from the late 19th Century and matured in the decade of 1930s. It is a major subject in its own right, and provides quantitative relations of internal energy, entropy etc. from atomistic considerations. However, in *common applications* in the metals and materials science and engineering, it is not widely employed. Hence, in this chapter, only introductory concepts and a few quantitative relations would be briefly presented, primary attention being devoted to interpretation of internal energy and entropy.

12.1 The Third Law of Thermodynamics

12.1.1 Nernst Heat Theorem

Based on experimental evidence, Nernst postulated that for chemical reaction between *pure solids* or *liquids,* at temperature approaching absolute zero (i.e. $T \rightarrow 0$),

$$\left[\frac{\delta(\Delta G^0)}{\delta T}\right]_P \rightarrow 0, \quad \left[\frac{\delta(\Delta H^0)}{\delta T}\right]_P \rightarrow 0 \tag{12.1}$$

Now,

$$\left[\frac{\delta(\Delta G^0)}{\delta T}\right]_P = -\Delta S^0$$

which is the same as Eq. (6.27). Hence, from Eq. (12.1) and (6.27),

$$\Delta S^0 \rightarrow 0 \quad \text{as} \quad T \rightarrow 0 \tag{12.2}$$

Again,

$$\Delta G^0 = \Delta H^0 - T\Delta S^0 \tag{6.3}$$

for isothermal processes. Differentiating Eq. (6.3) with respect to T at constant P, we get

$$\left[\frac{\delta(\Delta G^0)}{\delta T}\right]_P = \left[\frac{\delta(\Delta H^0)}{\delta T}\right]_P - \left\{\Delta S^\circ + T\left[\frac{\delta(\Delta S^0)}{\delta T}\right]_P\right\} \tag{12.3}$$

At $T \rightarrow 0$, combining Eqs. (12.1)–(12.3), we obtain

$$\left[\frac{\delta(\Delta S^0)}{\delta T}\right]_P \rightarrow 0 \tag{12.4}$$

Now, from Eq. (4.34),

$$T\left[\frac{\delta(\Delta S^0)}{\delta T}\right]_P = T\frac{\Delta C_P^0}{T} = \Delta C_P^0 \tag{12.5}$$

where

$$\Delta C_P^0 = \sum_{\text{product}} C_P^0 - \sum_{\text{reactant}} C_P^0 \tag{3.26}$$

Hence, from Eqs. (12.4) and (12.5), for reactions involving pure solids and liquids,

$$\Delta C_P^0 \rightarrow 0, \quad \text{as } T \rightarrow 0 \tag{12.6}$$

In accordance with Eq. (12.2), Nernst's Heat Theorem states that "for all reactions involving pure substances in condensed state, ΔS is zero at the absolute zero of temperature".

12.1.2 Entropy of a Pure Substance

A pure substance is either a pure element or a pure compound. Consider the formation of a compound AB from elements A and B, i.e.

$$A(s) + B(s) = AB(s) \tag{12.7}$$

for which

$$\Delta S^0 = S_{AB}^0 - S_A^0 - S_B^0 \tag{12.8}$$

At absolute zero temperature, $\Delta S^0 = 0$. It is possible provided either

(i) $S_{AB}^0 = S_A^0 + S_B^0$

or

(ii) S_{AB}^0, S_A^0, S_B^0 are all individually zero.

From a probability point of view, the first alternative is very unlikely as a general feature. It may be true, by chance, in a few cases. Hence, alternative (ii) is accepted as of general validity, and *it may be concluded that entropies of pure solids at T = 0 are zero*. Since all substances are solids at $T = 0$, we are omitting liquids. Therefore, the absolute value of entropy of a pure substance at any temperature can be determined by taking $T = 0$, $S^0 = 0$ as the lower limit. This is in contrast to energy, whose absolute value cannot be found out within the scope of conventional thermodynamics. This is the basis on which we find as to how thermochemical data tables provide values of entropy for pure substances. Figure 12.1 presents variation of S^0 with temperature for a few substances.

Figure 12.1 shows some increase in entropy during melting. In this connection, it is worthwhile to note down the following rules which were proposed long back from experimental observations.

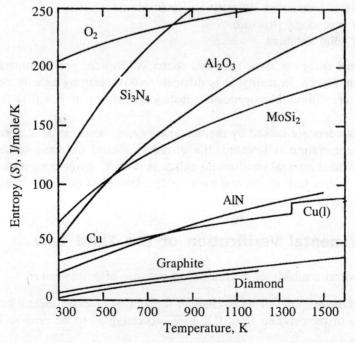

Fig. 12.1 Variation of molar entropies of some elements and compounds with temperature.

(i) *Richard's rule*: $\Delta S_m^0 \approx 10 - 20$ J K^{-1}mol^{-1}, for melting

(ii) *Trouton's rule*: $\Delta S_v^0 \approx 100$ J K^{-1}mol^{-1}, for vaporization

These rules are empirical guidelines and are approximate only. They have been explained from atomistic viewpoint. However, it is being omitted for the sake of conciseness.

12.1.3 Fundamental Significance of The Third Law

After Nernst, Max Planck modified Nernst's statement as follows:

"The entropy of any homogeneous substance, which is in complete internal equilibrium, may be taken as zero at 0 K".

As stated in section 4.4.2, one of the interpretations of entropy is that it is a measure of disorder in a substance. Its concrete proof is based on deductions in statistical thermodynamics, and will be presented in Section 12.3. Right now, we shall assume it to be correct and proceed further.

When disorder in a substance completely vanishes, it becomes *perfectly ordered and has only one possible arrangement of atoms*. Then and only then a complete (i.e. absolute) internal equilibrium is established. The causes for deviation from perfect order are:

1. Thermal energy and resulting motions of atoms and molecules
2. Noncrystallinity, as in amorphous solids and liquids
3. Crystal defects—vacancies, interstitials, dislocations
4. Disordered compound structure
5. Disordered solid solution.

At $T = 0$, thermal energy is zero. Thus this source of disorder gets eliminated. But other sources may remain. Hence, in reality, it is difficult to have entropy exactly zero at $T = 0$. However, for pure crystalline, homogeneous substances, entropy at $T = 0$ is fairly close to zero.

Actually, all disorders are caused by thermal energy and, hence, exist at normal and high temperatures. As temperature is lowered, the disorders should decrease continuously and vanish at $T = 0$, provided internal equilibrium exists. However, at low temperatures, the rates are very low. This causes lack of internal equilibrium. Disorders of higher temperatures, so to say, get 'frozen' at lower temperatures.

12.1.4 Experimental Verification of the Third Law

There are three general methods of evaluation of entropy of a substance:

(i) The *Third Law method.* This consists of integration of experimental heat capacities as function of temperature from 0 K to T, assuming S^0 to be zero at $T = 0$. From Eq. (4.38),

$$S_T^0 = \int_0^{T_m} \frac{C_P^0(s)}{T}\, dT + \Delta S_m^0 + \int_{T_m}^{T_b} \frac{C_P^0(l)}{T}\, dT + \Delta S_v^0 + \int_{T_v}^{T} \frac{C_P^0(g)}{T}\, dT \qquad (12.9)$$

This is the most general form of the equation which takes into account transformation of the solid into liquid and gas. If at temperature T, it remains a solid, then Eq. (12.9) gets simplified into:

$$S_T^0 = \int_0^{T} \frac{C_P^0(s)}{T}\, dT \qquad (12.10)$$

(ii) The *spectroscopic method*. In this method, we evaluate entropy of an ideal gas from spectroscopic data. The procedure for calculation is based on statistical mechanics.

(iii) The *Second Law method*. This consists of evaluation of ΔG^0 of reactions over a range of temperature. ΔS^0 is evaluated from variation of ΔG^0 with temperature on the basis of Eq. (6.27).

Comparison of predictions of these methods provided experimental verification of the Third Law. Comparison of results of all three methods is not common. One example is entropy of chlorine gas at room temperature (i.e. 298 K).

Method	S^0 per mol Cl_2, J mol^{-1}K^{-1}
Third Law	223.2
Spectroscopic	223.2
Second Law	221.9

The close agreement provides verification of the Third Law. It may be noted that the agreement between the Third Law and the spectroscopic method is good. Similar very close agreements have been found for several ideal gases by these two methods.

The oft-quoted illustrations of experimental verification of Third Law are through studies on the following allotropic transformations:

(i) Transformation of monoclinic sulphur into rhombohedral sulphur at 368.5 K
(ii) Transformation of white tin into gray tin at 292 K.

The procedure of verification based on the first case is illustrated in Fig. 12.2. Entropy is a state property and independent of path. But changes of entropy can be found out only by following reversible paths (see Section 4.3). For Fig. 12.2,

$$\sum_{\text{loop}} \Delta S^0 = 0 = \Delta S_{IV}^0 + \Delta S_{III}^0 - \Delta S_{II}^0 + \Delta S_{I}^0 \qquad (12.11)$$

So,

$$\Delta S_{IV}^0 = \Delta S_{II}^0 - (\Delta S_{III}^0 + \Delta S_{I}^0) \qquad (12.12)$$

The experimental values are:

$$\Delta S_{II}^0 = 1.093 \text{ J mol}^{-1}\text{K}^{-1}$$

$$\Delta S_{I}^0 + \Delta S_{III}^0 = 0.963 \text{ J mol}^{-1}\text{K}^{-1}$$

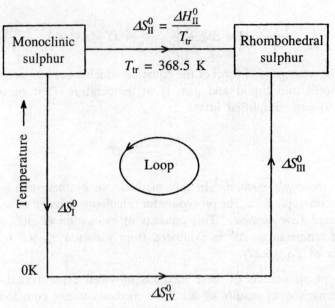

Fig. 12.2 An example of experimental verification of the Third Law of Thermodynamics.

Hence,

$$\Delta S_{IV}^0 = \text{entropy of transformation at } 0 \text{ K} = 1.093 - 0.963 = 0.13 \text{ J mol}^{-1} \text{ K}^{-1} \simeq 0$$

This is the verification of the Third Law within experimental errors.

12.2 Introduction to Statistical Thermodynamics

12.2.1 Kinetic Theory and Ideal Gas

In traditional mechanics, one form or the other of Newton's laws is employed to analyze the dynamics of system, and to follow motions of individual particles or elements (e.g. fluid elements) in a system as function of time. However, there are systems which consist of a large number of particles in motion and are too complex to analyze in terms of individual particles. These systems usually appear disordered because of the large number of particles involved, and the innumerable ways of sharing energy of the particles amongst themselves. Thermodynamics deals with such systems more easily.

The basic laws of thermodynamics deal with relationships amongst macroscopic variables, such as pressure, temperature, volume and internal energy. Such relations are simplest for ideal gases. These laws and relations do not, however, consider the behaviour of atoms and molecules which constitute a system.

The alternative approach to explain these macroscopic thermodynamic variables is the so-called Kinetic Theory. It developed gradually from the 17th Century and matured in the 19th Century. It was successfully applied to ideal gases on the postulate that the atoms/molecules of a gas are always in motion due to their thermal energy. The thermodynamic variables are related to this and are macroscopic averages. Newtonian mechanics is the basis of dealing with the velocities and kinetic energies of individual molecules.

The equations derived by the kinetic theory of ideal gases are the following:

$$\text{Pressure} = P = \frac{1}{3}\rho\overline{u^2} \qquad (12.13)$$

where ρ is density of gas and $\overline{u^2}$ is mean square (i.e. rms) of linear velocities of the molecules. Again,

$$\frac{1}{2}M\overline{u^2} = \frac{3}{2}RT \qquad (12.14)$$

where M is molecular mass. That is, the average translational kinetic energy per mole of an ideal gas is proportional to its temperature. Dividing Eq. (12.14) by Avogadro's number (N_0), we get

$$\frac{1}{2}m\overline{u^2} = \frac{3}{2}k_BT \qquad (12.15)$$

where m is mass of one molecule, and $k_B = R/N_0$ = Boltzmann's constant.

The internal energy of a monatomic ideal gas was assumed to be only due to translational motion of molecules. Hence, for 1 mole of gas, from Eq. (12.14),

$$U = \frac{3}{2}RT \qquad (12.16)$$

Classical statistical mechanics (section 12.2.2) arrived at the *Principle of Equipartition of Energy* amongst all independent degrees of freedom. In a monatomic gas, there are three degrees of freedom of translational motion along the three perpendicular coordinates. Hence, each is associated with energy of $1/2\,RT$ per mole. In a diatomic gas, there are five degrees of freedom (viz. three translational and two rotational). Hence,

$$U = 5 \times \frac{1}{2}RT = \frac{5}{2}RT \qquad (12.17)$$

12.2.2 Statistical Mechanics

Classical formulations of statistical mechanics were done in the late 19th century by Maxwell, Gibbs and Boltzmann. The kinetic theory predicted average translational kinetic energy of a gas. However, it could not predict distribution of speed and energy amongst molecules. We take recourse to statistical mechanics for this purpose. Besides mechanics, it is founded on the mathematical science of probability and statistics, an emrging field at that period.

Maxwell first solved the problem of the distribution of speeds in a gas containing a large number of molecules in a box. The Maxwellian speed distribution for a sample of gas at temperature T containing N molecules, each of mass m is

$$n(u) = 4\pi N \left(\frac{m}{2\pi k_B T} \right)^{3/2} u^2 \exp\left(-\frac{mu^2}{2k_B T} \right) \qquad (12.18)$$

where *n(u) is the probability distribution function of speed*, $n(u)$. du is the number of molecules in the gas sample having speed between u and $u + du$. Therefore,

$$N = \int_0^\infty n(u)\, du \qquad (12.19)$$

Figure 12.3 presents Maxwellian distribution of $n(u)$ as function of u at $T = 300$ K. The distribution of speeds of molecules in a liquid also resembles the curve in Fig. 12.3. However, very few molecules have large enough speed to break bond and escape as vapour molecules. It is also interesting to note that Maxwell's equation could be experimentally verified almost 100 years later (1955), when sufficiently precise experimental techniques could be employed.

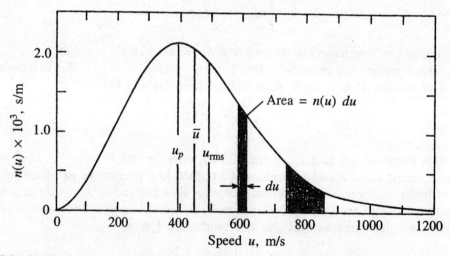

Fig. 12.3 Maxwellian speed distribution for the molecules of oxygen gas as function of u at $T = 300$ K. The number of molecules with speeds in any interval du is $n(u)du$. The number with speeds between u_1 and u_2 is given by the curve between these limits.

Following the derivation of speed distribution, energy distribution function was subsequently derived as

$$n(\varepsilon) = \frac{2N}{\sqrt{\pi}} \frac{1}{(k_B T)^{3/2}} \varepsilon^{1/2} \exp\left(-\frac{\varepsilon}{k_B T} \right) \qquad (12.20)$$

where $n(\varepsilon) \cdot d\varepsilon$ is number of molecules with energy between ε and $\varepsilon + d\varepsilon$.

$$U = \int_0^\infty n(\varepsilon)\, d\varepsilon \tag{12.21}$$

upon integration of Eq. (12.21) by combining it with Eq. (12.20), we get

$$U = \frac{3}{2} N_0\, k_B T = \frac{3}{2} RT \text{ for 1 mole of gas}$$

which is the same as Eq. (12.16) derived from the kinetic theory of gases.

12.2.3 Quantum Mechanics and Statistical Thermodynamics

Internal energy and energy distribution in an ideal gas were successfully predicted by statistical mechanics. Here the molecules were free and moved like distinct particles. However, in solids and liquids, the atoms/molecules are close and interacting. This gives rise to complex interaction and energy exchanges between them. In this, the electrons also participate. It required development of the subject of quantum mechanics to tackle such situations.

In the late 19th Century, there were two empirical laws of radiation, viz. Wein's Law and Rayleigh-Jean's Law. They could not be reconciled on the basis of the wave nature of radiation. In 1901, Planck reconciled them by postulating that radiation is emitted discretely as *energy packets*. It was a revolutionary concept. In 1905, Einstein called these packets as *quanta*. Thus Planck's theory came to be known as *Quantum Theory*.

The postulate of the quantum theory can be stated as follows: "If a particle is confined to move within a fixed volume, then its energy is quantized, i.e. discretized". Application of quantum theory later explained the structure of an atom by assuming that electrons can only be at some discrete energy levels.

On the basis of quantum theory and the techniques of statistical mechanics, the subjects of quantum mechanics and statistical thermodynamics were developed by several 20th Century physicists. An important postulate is that *all particles are exactly alike (i.e. indistinguishable from one another)*.

12.3 Basic Concepts and Relations in Statistical Thermodynamics

12.3.1 Basic Concepts and Postulates

In classical thermodynamics, we deal with *Macrostates* of substances. According to statistical interpretation of thermodynamics, *each macrostate consists of a very large number of Microstates* (also known as *quantum states*).

Chemical and physico-chemical equilibria are dynamic in nature, and are attained when the forward and backward processes have equal rates, thus balancing each other and resulting into zero net rate. There are always fluctuations in microscale. But these fluctuations are not perceptible for a large assembly of molecules. Macroscopic quantities (pressure, temperature, energy etc.) are only averages. Such averaging is to be done only statistically. For such statistical analysis, the basic postulates of statistical thermodynamics are as follows:

1. Each of the accessible and distinguishable quantum states (i.e. each microstate) *of a system of fixed energy* is equally probable.
2. The equilibrium state corresponds to the *most probable macrostate*. It is a state which contains the largest number of microstates.
3. As has already been stated, all particles are exactly alike.

Equation (12.20) provides the distribution function for energy of molecules in ideal gas. It is known as *Maxwell–Boltzmann (MB) statistics.*

On the basis of the above postulates (i.e. utilizing quantum theory), the following two energy distribution functions *at the most probable macrostate* were derived later by physicists whose names they bear

1. Bose–Einstein quantum statistics
2. Fermi–Dirac quantum statistics.

Each of these has played a very important role, not only in statistical thermodynamics but also in quantum mechanics quantum physics.

All the above three statistics are applicable to systems, where mass, composition, internal energy and volume are fixed. They differ in some other assumptions, the discussion of which is beyond the scope of this text.

12.3.2 The Most Probable Macrostate

As already stated, the most probable macrostate corresponds to the equilibrium state with which classical thermodynamics is concerned with. This state corresponds to the largest number of microstates, which means the *largest possible number of arrangements of particles (W) amongst the different discrete energy levels of the system.* In other words, at equilibrium, $W = W_{max}$.

The mathematical procedure of derivation of W_{max} is illustrated now. Let n_0, n_1, ..., n_i, ... n_r be the number of particles at energy levels ε_0, ε_1, ... ε_i, ..., ε_r. Then,

$$\text{Total No. of particles} = n_0 + n_1 + ... + n_i + ... + n_r = \sum_i n_i = N = \text{a constant}$$

$$(12.22)$$

From mathematics, the number of ways N particles can be arranged amongst the various energy levels are given as:

$$W = \frac{N!}{n_0! \, n_1! \, ... \, n_i! \, ... \, n_r!} = \frac{N!}{\Pi n_i!} \tag{12.23}$$

where '!' is the symbol for factorial.

Since a system consists of a large number of particles (i.e. atoms etc.), say of the order of Avogadro's number, *Stirling's approximation* is valid and, accordingly,

$$\ln W = (N \ln N - N) - \sum_i (n_i \ln n_i - n_i) \tag{12.24}$$

or

$$d(\ln W) = -\sum_i \left(\ln n_i \, dn_i + \frac{n_i}{n_i} \, dn_i - dn_i \right) = -\sum_i \ln n_i \, dn_i \tag{12.25}$$

since N = a constant. Again,

$$d(\ln W) = 0 \quad \text{at } W = W_{max} \tag{12.26}$$

12.3.3 Internal Energy and Entropy at Most Probable Macrostate

Internal energy of the system of section 12.3.2 is given as

$$U = n_0\varepsilon_0 + n_1\varepsilon_1 + \ldots + n_i\varepsilon_i + \ldots + n_r\varepsilon_r = \sum_i n_i\varepsilon_i \tag{12.27}$$

Since U is assumed to be a constant,

$$dU = \sum_i \varepsilon_i \, dn_i + \sum_i n_i \, d\varepsilon_i = 0 \tag{12.28}$$

Since $\varepsilon_0, \varepsilon_1 \ldots, \varepsilon_i, \ldots \varepsilon_r$ are all constants (i.e. fixed discrete energy levels as per quantum theory)

$$\sum_i n_i \, d\varepsilon_i = 0 \tag{12.29}$$

Combining Eqs. (12.28) and (12.29), we get

$$dU = \sum_i \varepsilon_i \, dn_i = 0 \tag{12.30}$$

Again from Eq. (12.22), since N is a constant,

$$dN = \sum_i dn_i = 0 \tag{12.31}$$

Combining Eqs. (12.25), (12.26), (12.30) and (12.31), we may write

$$\sum_i (\ln n_i + \alpha + \beta\varepsilon_i) \, dn_i = 0 \tag{12.32}$$

where α and β are *multipliers*.

Equation (12.32) is satisfied by substituting the following relation into it (details skipped) *at the most probable macrostate (i.e. equilibrium state):*

$$n_i = \frac{N \exp(-\beta\varepsilon_i)}{P} \tag{12.33}$$

where

$$\beta = \frac{1}{k_B T} \tag{12.34}$$

$$P = \text{partition function} = \sum_i \exp(-\beta \varepsilon_i) \tag{12.35}$$

The above equations give a distribution function of n_i as proportional to $\exp(-\varepsilon_i/k_B T)$, and is analogous to Maxwell–Boltzmann distribution function [Eq. (12.20)].

It can be derived further (derivation skipped) that, at the most probable macrostate,

$$\ln W = N \ln P + \frac{U}{k_B T} \tag{12.36}$$

or

$$d(\ln W) = \frac{dU}{k_B T} \tag{12.37}$$

since both N and P are constants.

At constant volume, from Eq. (2.18),

$$dU = \delta q \tag{12.38}$$

where q is heat flow from surrounding into the system. Hence,

$$d(\ln W) = \frac{\delta q}{k_B T} \tag{12.39}$$

At constant temperature and for reversible processes, from Eq. (4.31),

$$\frac{\delta q}{T} = dS \tag{12.40}$$

Combining Eqs. (12.39) and (12.40) and integrating, we get

$$S = k_B \ln W + \text{constant} \tag{12.41}$$

From the Third Law of Thermodynamics, $S = 0$ at $T = 0$ for a perfectly ordered solid (see section 12.1.3). As mentioned there, perfect order means only one unique arrangement of particle in the system, i.e. $W = 1$ (i.e. $\ln W = 0$). Therefore, the constant in Eq. (12.41) drops out and,

$$S = k_B \ln W \tag{12.42}$$

This equation was proposed by Planck in 1906, and is known as *Boltzmann equation*. It relates entropy with the number of arrangements of particles amongst different energy levels available in a system of fixed mass, volume, number of particles, as also of fixed internal energy and composition.

12.3.4 Entropy of Mixing for a Binary Solution

Consider mixing of atoms A and B to form a binary solution. The process may be represented as

$$A + B \text{ (unmixed)} \rightarrow A + B \text{ (mixed)} \qquad (12.43)$$
$$\text{(state 1)} \qquad\qquad \text{(state 2)}$$

Integral molar entropy of mixing is

$$\Delta S^m = S_2 - S_1 = (S_{th} + S_{conf})_2 - (S_{th} + S_{conf})_1 \qquad (12.44)$$

where subscripts 2 and 1 denote state 2 and 1, respectively.

Entropy is a measure of disorder, qualitatively speaking, and such disorder in the solution primarily originates from

(i) motion and vibration of atoms due to their thermal energy, entropy from this source being known as *thermal entropy* (S_{th}); and
(ii) arrangements of atoms in various lattice sites, entropy from this source being known as *configurational entropy* (S_{conf}).

Assuming $(S_{th})_2 = (S_{th})_1$, combination of Eq. (12.44) with Eq. (12.42) yields

$$\Delta S^m = (S_{conf})_2 - (S_{conf})_1 = k_B[\ln (W_{conf,2}) - \ln (W_{conf,1})] \qquad (12.45)$$

In unmixed state, configurations of both A and B are unique since they exist separately. Hence, $W_{conf,1} = 1$ and, therefore,

$$\Delta S^m = k_B \ln (W_{conf,2}) \qquad (12.46)$$

Let us take 1 mole of solution. The total number of atoms is Avogadro's number (N_0). Assume the total number of sites = N_0, and also assume the number of A and B atoms as n_A and n_B, respectively. From the theory of combination, n_A atoms can be arranged in N_0 sites in $N_{0}C_{n_A}$ ways assuming random mixing of A and B atoms. Once A atoms are arranged, B atoms occupy the remaining sites (i.e. one arrangement only). Therefore,

$$W_{conf,2} = N_{0}C_{n_A} \cdot 1 = \frac{N_0!}{n_A!\,(N_0 - n_A)!} = \frac{N_0!}{n_A!\,n_B!} \qquad (12.47)$$

From Eqs. (12.46) and (12.47),

$$\Delta S^m = k_B \ln \left[\frac{N_0!}{n_A!\,n_B!}\right] \qquad (12.48)$$

Employing Stirling's approximation,

$$\Delta S^m = k_B[(N_0 \ln N_0 - N_0) - (n_A \ln n_A - n_A) - (n_B \ln n_B - n_B)] \qquad (12.49)$$

Noting that $n_A + n_B = N_0$, rearrangement of Eq. (12.49) leads to

$$\Delta S^m = k_B \left[n_A \ln \left(\frac{N_0}{n_A} \right) + n_B \ln \left(\frac{N_0}{n_B} \right) \right] \tag{12.50}$$

Since n_A/N_0 and n_B/N_0 are mole fractions of A and B (i.e. X_A and X_B), and $k_B N_0 = R$, Eq. (12.50) may be rewritten as

$$\Delta S^m = - R(X_A \ln X_A + X_B \ln X_B) \tag{12.51}$$

Equation (12.51) is identical with Eq. (9.52) for ideal binary solution. Hence, it may be concluded that an ideal solution is characterized by

    (i) random mixing of atoms/molecules, and
    (ii) no change in thermal entropy when the solution forms from pure components.

12.4   Summary

1. Max Planck stated the Third Law of Thermodynamics as

   The entropy of any homogeneous substance, which is in complete internal equilibrium, may be taken as zero at 0 K.

   This is valid only for a perfectly ordered, pure, crystalline solid.

2. The Third Law was originally proposed on the basis of experimental measurements at temperatures close to 0 K. It was subsequently verified for several substances systematically from experimental data.

3. For ideal gases, the following equations were derived on the basis of the kinetic theory:

$$\text{Pressure } P = 1/3 \, \rho \overline{u^2}$$

   where ρ is density of gas, and $\overline{u^2}$ is the rms of linear velocities of molecules.

   $U = 3/2RT$, for a monatomic gas due to translational motion of molecules.

4. Subsequently, distribution functions for speed and energy of molecules in ideal gas were successfully derived on the basis of statistical mechanics.

5. In condensed phases, atoms/molecules are close and mutually interacting, in contrast to ideal gases where they are free. Energy distributions could be derived in these only after the advent of the Quantum Theory. The subjects of Quantum Mechanics and Statistical Thermodynamics, then, were developed on the basis of statistical mechanics and quantum theory.

6. Some basic postulates and relations of statistical thermodynamics, such as microstate, macrostate, most probable macrostate etc., have been briefly presented in this chapter.

7. At the most probable macrostate, which corresponds to equilibrium state in classical thermodynamics, it has been derived that:

   (i) $n_i = \dfrac{N \exp(-\beta \varepsilon_i)}{P}$

   where $\beta = 1/k_B T$, P = partition function = $\sum_i \exp(-\beta \varepsilon_i)$,

   where n_i is number of particles at energy level ε_i, and N is total number of particles.

   (ii) $S = k_B \ln W$

   where W is number of ways particles in an assemblage can be arranged.

8. Finally, entropy of mixing of a binary solution A-B has been derived on the basis of above as

$$\Delta S^m = -R(X_A \ln X_A + X_B \ln X_B)$$

This is the same as that of an ideal solution. Assumptions in derivation are random mixing, and no change of thermal entropy upon mixing.

Chapter *13*

Thermodynamics of Electrochemical Cells

13.1 Introduction

An electrochemical reaction involves coupling of chemical reaction with flow of electric current. It is a very important field in metallurgy and materials science, as the following examples will illustrate:

- Many metals are extracted and/or refined by electrolytic process (Zn, Al, Mg etc.).
- Electroplating and anodizing are widely employed for surface protection of metals and alloys from corrosion.
- Electrochemical reactions occur in corrosion, hydrometallurgy, and slag-metal reactions.
- Batteries are electrochemical cells.
- The electrochemical method is an important tool for thermodynamic measurements, especially at high temperatures.

Electrochemical cells are broadly classified into:

(i) *Galvanic cells (i.e. batteries),* where stored chemical energy is converted into electrical energy
(ii) *Electrolytic cells,* i.e. cells for electrolysis, where electrical energy is used to do chemical work.

Thermodynamic studies/predictions/measurements can be properly done only for galvanic cells since these can be made to operate reversibly. It may be recalled that the term 'galvanic' originated from the famous Italian scientist Galvani, who was a pioneer in this field.

13.1.1 The Daniel Cell

The most common example of a galvanic cell is a Daniel cell, shown in Fig. 13.1.

Overall reaction: $Zn(s) + CuSO_4(aq) = Cu(s) + ZnSO_4(aq)$ (13.1)

where "aq" means aqueous solution.

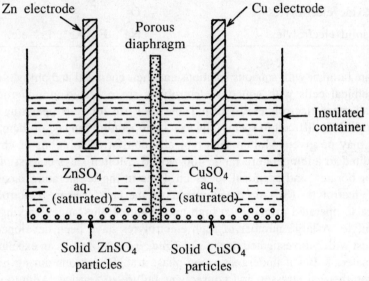

Fig. 13.1 Sketch of a Daniel cell.

Figure 13.1 shows that both $CuSO_4$ and $ZnSO_4$ solutions are *saturated* in a Daniel cell. The overall reaction may also be written in ionic form as

$$Zn(s) + Cu^{2+}(aq) = Cu(s) + Zn^{2+}(aq)$$ (13.2)

Reaction (13.2) consists of:

Anodic reaction: $Zn(s) = Zn^{2+}(aq) + 2e^-$ (13.3)

Cathodic reaction: $Cu^{2+}(aq) + 2e^- = Cu(s)$ (13.4)

The aqueous solutions constitute the electrolytes in the Daniel cell. Electrolytes are ionic liquids or solids and may be classified as given in Table 13.1, along with examples of some ions, which are mobile and carry electric current in the electrolytes.

It is to be emphasized here that an electrolyte may have several current carrying ions, but not all of them participate in electrochemical reaction. For example, during electrolysis of aqueous $ZnSO_4$ solution, some H_2SO_4 is added to improve its electrical conductivity. Although H^+ will carry the major fraction of current through the aqueous solution due to its high mobility, the reaction at cathode would be primarily discharge of Zn^{2+} and deposition of Zn. Similarly, a molten salt at high temperature may consist of NaCl, KCl and $CdCl_2$, but it is only Cd^{2+} which will participate in electrochemical reaction.

Table 13.1   Classification of Electrolytes

Electrolytes	*Examples of current carrying ions*
Aqueous solutions	Na^+, H^+, Cl^- etc.
Molten salts	Na^+, K^+, Cd^{2+}, Cl^-, F^- etc.
Molten slags	Ca^{2+}, Mn^{2+} etc.
Solid oxide electrolytes	O^{2-}
Other solid electrolytes	Na^+, F^-, Ag^+, Li^+ etc.

You are familiar with aqueous solutions and their chemical and physico-chemical behaviour. Electrochemical cells with aqueous electrolytes are operated around room temperature. In contrast, all other electrolytes listed in Table 13.1 are for cells operating primarily at higher temperatures. It is difficult to specify the exact temperature ranges for them. Some approximate idea only may be given. Molten salts typically consist of solutions of chlorides or fluorides and are suited to a temperature range 400–900°C. Molten slags consist of solutions of metal silicates or borates, and are ideal for electrolytes in the temperature range of 1000–1500°C.

Solid electrolytes (SE) are developments in the last 50 years. Electrochemical cells with these are also operated primarily at high temperatures, the ranges varying with the nature of the electrolyte. A large number of such electrolytes have been developed. The best known and the most widely investigated one is *stabilized zirconia*. ZrO_2 is an excellent high temperature ceramic material. But it undergoes some phase transformations during heating and cooling. This causes thermal stresses and consequent failure in service. Addition of oxides such as CaO, MgO, and Y_2O_3 stabilizes the cubic fluorite (i.e. CaF_2) structure from room temperature upto its melting temperature (approx. 2400°C) and thus prevents failure.

Let us take the example of ZrO_2–CaO solid solution. In pure ZrO_2, Zr^{4+} ions occupy cationic lattice sites and O^{2-} ions anionic sites. On addition of CaO, divalent Ca^{2+} occupies some cationic sites in place of tetravalent Zr^{4+}. For maintainance of local electrical neutrality, replacement of a Zr^{4+} ion by one Ca^{2+} ion leads to removal of one O^{2-} from anionic site. This creates an oxygen ion vacancy and allows movement of O^{2-} ion by vacancy mechanism. Thus, O^{2-} is the only current carrying species since Zr^{4+} and Ca^{2+} are almost immobile. However, since the material is solid, a reasonably large mobility of O^{2-} can be attained only above 700–800°C. Thus the cell is to be operated only at high temperature.

For thermodynamic measurements, ZrO_2–CaO solid electrolyte has been widely used. Other important ones are molten salts·based on KCl and LiCl mixture as base electrolyte (as H_2O is base for aqueous solutions), and SEs such as ZrO_2-Y_2O_3, ThO_2-Y_2O_3, CaF_2 doped with YF_3.

13.2 Thermodynamics of Reversible Galvanic Cells

In practice, the same electrochemical cell can be made to behave either as a galvanic cell or as an electrolytic cell. Let us take the example of Daniel cell (Fig. 13.1). Let us connect it to an external DC source such that the externally imposed voltage (V_{ext}) opposes that of the Daniel cell (V_{cell}). If $V_{ext} < V_{cell}$, then the cell behaves as a galvanic cell with reaction (13.1) occurring as shown. However, if $V_{ext} > V_{cell}$, then the direction of reaction (13.1) is reversed, i.e. the reaction becomes

$$Cu(s) + ZnSO_4(aq) = Zn(s) + CuSO_4(aq) \qquad (13.5)$$

In this situation, external electrical energy is being consumed to force reaction (13.5) to take place, and the cell behaves as an electrolytic cell.

In Section 2.3, we have discussed the concept of a reversible process. We may recall that (a) it is very slow and (b) it occurs near equilibrium, and the process can be reversed by shifting it from equilibrium in the other direction.

In an electrochemical cell, chemical reaction occurs only if the current is allowed to flow through the cell. Ideally, no current flows and, hence, we have cell equilibrium, if the circuit is open. The voltage at this condition is known as *Electromotive force* (i.e. emf) *of a galvanic cell*. However, this emf can not be measured without a voltmeter. Use of a voltmeter of very high resistance makes the current flow negligible and, hence the cell behaves as a reversible galvanic cell. Another method is to impose voltage from an external source such that V_{ext} is very close to the cell emf, and current flow is negligible.

13.2.1 Relation between Cell EMF(E) and Free Energy of Cell Reaction (ΔG)

For a reversible process, from Section 5.1,

$$dG = V\,dP - S\,dT - \delta W' \qquad (5.10)$$

where W' is any form of work other than that done against pressure. In this chapter, W' would refer to *electrical work* performed by the galvanic cell.

At constant temperature and pressure,

$$(dG)_{T,P} = -\delta W', \text{ i.e. } (\Delta G)_{T,P} = -W' \qquad (5.48)$$

As discussed in Section 5.5, here W' is also known as maximum work, since work is maximum for a reversible process only.

Since in a reversible galvanic cell the voltage across the cell is its EMF, we have

$$\delta W' = E\delta q \qquad (13.6)$$

where E = cell emf and δq = infinitesimal quantity of electrical charge transferred across the cell due to chemical reaction in the cell.

Again, from *Faraday's Laws of electrolysis*,

$$\delta q = ZF\,dn \qquad (13.7)$$

where

   Z = valency involved in the chemical reaction

F = *Faraday's constant* = 96,520 joules/volt/gm-equivalent

dn = an infinitesimally small quantity of species (in gm.moles) electrochemically transferred between electrode and electrolyte as a result of δq quantity of charge.

Combining Eq. (5.48) with Eqs. (13.6) and (13.7), at constant T and P, we get

$$dG = -\delta W' = -ZFE\, dn \tag{13.8}$$

For a finite process, if n = 1 gm-mole, then

$$\Delta G = -ZFE \tag{13.9}$$

If the species transferred between electrodes and the electrolyte are at *their respective standard states*, then we write Eq. (13.9) as

$$\Delta G^0 = -ZFE^0 \tag{13.10}$$

where E^0 is known as the *standard emf* of the cell.

Again, from Section 7.3,

$$\Delta G = \Delta G^0 + RT \ln J \tag{7.24}$$

where

J = activity quotient.

From Eqs. (13.9), (13.10) and (7.24),

$$E = E^0 - \frac{RT}{ZF} \ln J \tag{13.11}$$

13.2.2 Sign Convention for EMF

We need not sketch an electrochemical cell in order to describe it. There is a standard convention for its short representation. Accordingly, for example, the Daniel cell may be represented as

$$\text{Zn(s)} \mid \text{ZnSO}_4\text{(aq)} \parallel \text{CuSO}_4\text{(aq)} \mid \text{Cu(s)} \tag{13.12}$$
$$\text{(sat)} \qquad\qquad \text{(sat)}$$

Single vertical lines separate electrode from electrolyte. The double vertical line means porous diaphragm.

An important question is: What should be the sign of the emf of the above cell-positive or negative? For this, some convention is required. There is the European convention and also the US convention. We shall adopt the *US convention. Accordingly, if the reaction proceeds spontaneously from left to right, the sign of E is positive.* For the Daniel cell, reaction proceeds spontaneously in the forward direction of Eq. (13.1) if the circuit is closed, i.e. Zn dissolves as ZnSO_4 and Cu is deposited at cathode from CuSO_4. Therefore, in the representation of Daniel cell (13.12), reaction proceeds spontaneously from left to right, and hence the emf is positive.

13.2.3 Cell EMF and Chemical Potentials at Electrodes

In Section 10.1, we have discussed chemical potential. It has presented the general statement of equality of chemical potentials of components in phases at equilibrium. If there are only two phases (I and II) at equilibrium, then at constant T and P,

$$\mu_1^I = \mu_1^{II}, \ \mu_2^I = \mu_2^{II}, \ ..., \ \mu_i^I = \mu_i^{II}, \ ... \tag{10.18}$$

where the subscripts denote components in solution (elements or compounds).

In an electrochemical cell, there is electrical work also. Let us consider the two electrodes as phases I and II. Then each electrode will have not only chemical potential, but also electrical potential. Let us consider exchange of component i between electrodes I and II. Then Eq. (10.14) is to be modified as follows:

$$dG'^{,\,I} = \mu_i^I \ dn_i^I + ZF\phi^I \ dn_i^I = (\mu_i^I + ZF\phi^I) \ dn_i^I \tag{13.13}$$

$$dG'^{,\,II} = \mu_i^{II} \ dn_i^{II} + ZF\phi^{II} \ dn_i^{II} = (\mu_i^{II} + ZF\phi^{II}) \ dn_i^{II} \tag{13.14}$$

ϕ^I and ϕ^{II} are electrical potentials of electrodes I and II respectively. $(ZF\phi) \ dn_i$ represents the electrical work to be performed for transfer of dn_i moles of component i. The quantity $(\mu_i + ZF\phi)$ is the *sum of chemical and electrical potential energy per mole of i,* and is known as the *electrochemical potential of component i.*

Proceeding further as in Chapter 10, for exchange of component i between the electrodes, we have

$$dn_i^{II} = - \ dn_i^I = - \ dn_i \tag{13.15}$$

$$dG' = dG'^{,\,I} + dG'^{,\,II} = dn_i \left[(\mu_i^I + ZF\phi^I) - (\mu_i^{II} + ZF\phi^{II}) \right] \tag{13.16}$$

The two electrodes are at *electrochemical equilibrium* with one another *(not chemical equilibrium)* when there is no current flow (i.e. cell voltage = emf = E). Under this situation, $dG' = 0$. Therefore, from Eq. (13.16), at constant T and P,

$$\mu_i^I + ZF\phi^I = \mu_i^{II} + ZF\phi^{II} \tag{13.17}$$

In other words, *the electrochemical potentials are equal at the two electrodes for reversible exchange of component i between the electrodes.* Again, under this condition,

$$\phi^{II} - \phi^I = \text{cell emf} = E \tag{13.18}$$

Combining Eqs. (13.17) and (13.18), we get

$$\mu_i^{II} - \mu_i^I = -ZF(\phi^{II} - \phi^I) = -ZFE \tag{13.19}$$

Equation (13.9) relating E with ΔG is of general applicability to any emf cell. Equation (13.19) is applicable only to concentration cells. Examples of both will be presented in the following section.

13.2.4 Temperature Coefficient of EMF

Let us recall the following equations from Sections 6.2 and 6.3,

$$\Delta G = \Delta H - T\Delta S \tag{6.4}$$

for an *isothermal process*. Also,

$$\left[\frac{\delta(\Delta G)}{\delta T}\right]_P = -\Delta S \tag{6.27}$$

Combining the above equations with Eq. (13.9), we obtain

$$\Delta S = ZF\left(\frac{\delta E}{\delta T}\right)_P \tag{13.20}$$

$$\Delta H = -ZFE + ZFT\left(\frac{\delta E}{\delta T}\right)_P \tag{13.21}$$

13.3 Examples of EMF Cells

Electrochemical cells may be classified into:

  (i) *reaction cells,* where the source of cell emf is a chemical reaction,
  (ii) *concentration cells*, where the source of emf is the difference in concentration of a component i at the two electrodes.

13.3.1 Examples of Reaction Cell

EXAMPLE 13.1 Find out E^0 for the Daniel cell at 298 K. Also express E as function of activities if the electrolytes are not saturated solutions.

Given: ΔG^0 for reaction (13.1) is –213.0 kJ mol^{-1} at 298 K.

Solution From Eq. (13.10),

$$E^0 = -\frac{\Delta G^0}{ZF} = -\frac{(-213 \times 10^3)}{2 \times 96520} = 1.104 \text{ volts} \tag{E.13.1}$$

From Eq. (13.11),

$$E = 1.104 - \frac{8.314 \times 298}{2 \times 96500} \ln J = 1.104 - 0.0128 \ln J \tag{E.13.2}$$

where

$$J = \frac{[a_{Cu}]\,(a_{ZnSO_4})}{[a_{Zn}]\,(a_{CuSO_4})} \tag{E.13.3}$$

a_{Cu} and a_{Zn} are activities of Cu and Zn at respective electrodes. Since they are pure, $a_{Cu} = a_{Zn} = 1$. In a Daniel cell, the electrolytes are saturated solutions. It means that the $ZnSO_4$ solution co-exists at equilibrium with solid $ZnSO_4$, and the $CuSO_4$ solution with solid $CuSO_4$. In other words, in Daniel cell, the activities of $ZnSO_4$ and $CuSO_4$ in the electrolytes are also 1. However, in unsaturated electrolytes, the activities in electrolytes will be less than 1. Hence, from Eqs. (E.13.2) and (E.13.3), at 298 K,

$$E = 1.104 - 0.0128 \ \ln \frac{(a_{ZnSO_4})}{(a_{CuSO_4})} \tag{E.13.4}$$

EXAMPLE 13.2 At 1300 K, the emf of the following cell is 0.60 V.

$$\begin{array}{ccc} Pb & | \quad Liquid \quad | & Pt(s), \ O_2(g) \\ (pure \ liquid) & PbO - SiO_2 & (p_{O_2} = 1 \ atm) \end{array} \tag{E.13.5}$$

$$\begin{array}{c} electrolyte \\ (X_{PbO} = 0.5) \end{array}$$

Given: $\Delta G^0 = -88{,}800 \ J \ mol^{-1}$ at 1300 K for the reaction

$$Pb(l) + 1/2 \ O_2(g) = (PbO) \tag{E.13.6}$$

Calculate the activity and the activity coefficient of PbO in the $PbO\text{-}SiO_2$ electrolyte.

Solution In the above cell, solid platinum serves as the oxygen electrode since its surface comes to equilibrium with the gas phase quickly without forming oxide. The cell reaction is reaction (E.13.6), i.e. formation of PbO from Pb and O_2.
For the emf cell (E.13.5), therefore,

$$E_5 = E_5^0 - \frac{RT}{ZF} \ \ln J = E_5^0 - \frac{RT}{ZF} \ \ln \left[\frac{(a_{PbO})}{[a_{Pb}] \ p_{O_2}^{1/2}} \right] \tag{E.13.7}$$

Now, $E_5^0 = -\Delta G^0/ZF$, where $Z = 2$ for reaction (E.13.6). Hence,

$$E_5^0 = -\frac{(-88{,}800)}{2 \times 96500} = +0.460 \ volt$$

$a_{Pb} = 1$ since Pb is pure and $P_{O_2} = 1$. Hence,

$$E_5 = 0.460 - 0.056 \ \ln \ (a_{PbO}) \tag{E.13.8}$$

Since $E_5 = 0.60$ volts, we have

$$a_{PbO} = 0.083; \quad \gamma_{PbO} = a_{PbO}/X_{PbO} = 0.166 \ \text{[Ans.]}$$

13.3.2 Examples of Concentration Cell

EXAMPLE 13.3   Chlorides, fluorides etc. are low melting solids. Hence, for thermodynamic measurements, emf cells, with these molten salts as electrolytes, can be fabricated with glass. An example is the following cell:

$$\text{W} \mid \text{Cd (pure liquid)} \mid \text{CdCl}_2 \text{ dissolved in} \mid \text{[Cd-Pb]} \mid \text{W}$$

(electrode I)   molten LiCl-KCl   (liquid alloy)   (E.13.9)
                     solution as   (electrode II)
                     electrolyte

Here, W means tungsten lead wires. Calculate the activity of Cd in Cd-Pb alloy at 800 K, if the cell EMF is 0.1 volt.

Solution   It is a concentration cell. The emf is due to difference in concentration of cadmium at the two electrodes. Electrochemical exchange of Cd between the two electrodes occurs via Cd^{2+} ion in electrolyte. Tungsten serves as solid lead wire for connection with the outside circuit. Employing Eq. (13.19), we may write

$$\mu_{Cd}(\text{alloy}) - \mu^{\circ}_{Cd}(\text{pure Cd}) = -ZFE \qquad \text{(E.13.10)}$$

Again, from Eq. (10.25),

$$\mu_{Cd}(\text{alloy}) - \mu^{\circ}_{Cd} = RT \ln [a_{Cd}]_{\text{alloy}} \qquad \text{(E.13.11)}$$

Combining the above equations and noting that $Z = 2$, we obtain

$$E = -\frac{RT}{2F} \ln [a_{Cd}]_{\text{alloy}} \qquad \text{(E.13.12)}$$

Putting $E = 0.1$ V and $T = 800$ K, we get

$$a_{Cd} \text{ in alloy} = 0.055 \quad \text{(Ans.)}$$

We can also employ Eq.(13.11). E^0 is emf of a hypothetical cell where cadmium at standard state (i.e. pure Cd) constitutes both electrodes. Obviously then, $E^0 = 0$ for this system.

Again, the cell reaction (*actual process*) is

$$\text{Cd(pure)} = [\text{Cd}]_{\text{alloy}} \qquad \text{(E.13.13)}$$

For this, ΔG = partial molar free energy of mixing of Cd in the alloy = $\bar{G}^m_{Cd}$

$$J = \frac{[a_{Cd}]_{\text{alloy}}}{[a_{Cd}]_{\text{pure}}} = [a_{Cd}]_{\text{alloy}}$$

Combining all these, Eq. (13.11) yields

$$-2EF = \Delta G = \bar{G}^m_{Cd} = RT \ln [a_{Cd}]_{\text{alloy}} \qquad \text{(E.13.14)}$$

which is the same as Eq. (E.13.12). Therefore, this example demonstrates how either Eq. (13.19) or Eq. (13.11) can be employed for solving problems of concentration cells.

13.3.3 EMF Cells with Solid Oxide Electrolyte

In the solid oxide electrolyte, O^{2-} is the current carrying ion. The emf of such a cell is due to difference in chemical potentials of oxygen in the two electrodes. Hence, it is a concentration cell, with concentration difference of oxygen of the electrodes being the cause of emf.

A general representation of these cells with ZrO_2-CaO electrolyte is

$$O_2(g, \text{ at } p_{O_2}^I), \text{ Pt} \quad | \quad ZrO_2\text{-CaO} \quad | \quad \text{Pt}, O_2(g, \text{ at } p_{O_2}^{II}),$$
$$\text{(electrode I)} \quad \text{electrolyte } (O^{2-}) \quad \text{(electrode II)} \qquad \text{(E.13.15)}$$

On the basis of Eq. (13.19),

$$\mu_{O_2}^{II} - \mu_{O_2}^{I} = -ZFE \qquad \text{(E.13.16)}$$

Here $Z = 4$ since O_2 molecule has a valency of 4. Again, from Chapter 10,

$$\mu_{O_2} = RT \ln p_{O_2} \qquad \text{(10.27)}$$

Combining Eqs. (E.13.16) and (10.27), we get

$$E = \frac{\mu_{O_2}^I - \mu_{O_2}^{II}}{4F} = \frac{RT}{4F} \ln \frac{p_{O_2}^I}{p_{O_2}^{II}} \qquad \text{(E.13.17)}$$

Such cells have found widespread use in industry for measurement of oxygen dissolved in liquid steel, copper, furnace atmosphere control, combustion control and so on. These emf cells have also been and are still being used widely for precise thermodynamic measurements at high temperature. In fact, such cells act as *oxygen meters* (i.e. *oxygen sensors*) with capabilities to measure oxygen concentration or partial pressure in a wide range of temperature and p_{O_2}. These will be illustrated by the following examples.

EXAMPLE 13.4 Roy Chowdhury and Ghosh (*Metallurgical Transactions*, **2**, 2171–74, 1971) made thermodynamic measurements in liquid tin-silver alloys using solid electrolyte cell as follows:

$$\text{Cermet} \quad | \quad \text{Sn(l)} + SnO_2(s) \quad | \quad ZrO_2\text{–CaO} \quad | \quad \text{Sn–Ag(l)} + SnO_2(s) \quad | \quad \text{Cermet}$$
$$\text{(electrode I, } p_{O_2} = p_{O_2}^I) \text{ electrolyte} \quad \text{(electrode II, } p_{O_2} = p_{O_2}^{II}) \qquad \text{(E.13.18)}$$

The emf cell is shown in Fig. E.13.1. The apparatus and technique had been employed by Ghosh and co-workers earlier. The electrolyte was in the form of a crucible. Pure Sn melt with SnO_2 was in outer alumina crucible. The alloy melt with SnO_2 was inside the zirconia crucible. The cermets served as nonreacting leads, and the entire assembly was at constant temperature inside a furnace.

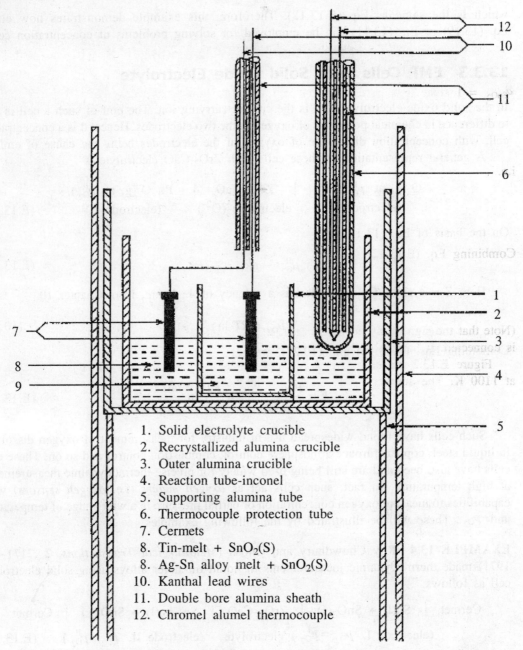

Fig. E.13.1 EMF cell for thermodynamic measurements in molten tin-silver alloys.

1. Solid electrolyte crucible
2. Recrystallized alumina crucible
3. Outer alumina crucible
4. Reaction tube-inconel
5. Supporting alumina tube
6. Thermocouple protection tube
7. Cermets
8. Tin-melt + $SnO_2(S)$
8. Ag-Sn alloy melt + $SnO_2(S)$
10. Kanthal lead wires
11. Double bore alumina sheath
12. Chromel alumel thermocouple

Tin is much more reactive than silver. Hence, the chemical reaction in both electrodes is

$$[Sn] + O_2(g) = SnO_2(s) \qquad (E.13.19)$$

Sufficient time was allowed for attainment of this equilibrium at each electrode, and

$$\text{Equilibrium constant } K = \left\{ \frac{(a_{SnO_2})}{[a_{Sn}]\, p_{O_2}} \right\}_{eq} \tag{E.13.20}$$

$a_{SnO_2} = 1$ since it is pure solid, $a_{Sn} = 1$ for pure liquid tin. Hence, from Eq. (E.13.20),

$$K = \frac{1}{p_{O_2}^{I}} = \frac{1}{[a_{Sn}]\, p_{O_2}^{II}} \tag{E.13.21}$$

i.e.

$$\frac{p_{O_2}^{I}}{p_{O_2}^{II}} = [a_{Sn}]_{alloy} \tag{E.13.22}$$

Combining Eq. (E.13.22) with Eq. (E.13.17), we get

$$E = \frac{RT}{4F} \ln [a_{Sn}]_{alloy} \tag{E.13.23}$$

(Note that the sign of E is negative. The sign of measured E depends on how the voltmeter is connected. It has to be taken as negative for calculation purposes.)

Figure E.13.2 presents activity vs. composition relationship in liquid Sn-Ag solution at 1100 K. The activity of tin was experimentally measured. The activity of silver was

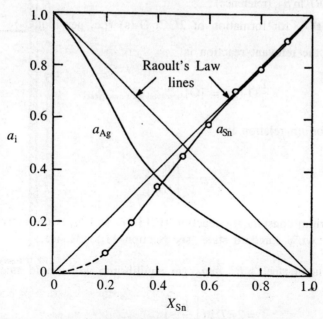

Fig. E.13.2 Activity-composition plots for liquid tin-silver alloys at 1100 K [P.J. Roy Chowdhury and A. Ghosh, *Met. Trans*, **2**, 2171, 1971].

calculated from experimental data by the Gibbs–Duhem integration using Darken's method (see Section 9.6.3). Pure silver is a solid at 1100 K. For GD integration, hypothetical liquid Ag standard state was employed. The procedure and equations for this conversion have been presented in section 10.3.3.

EXAMPLE 13.5 During steel making, measurement of dissolved oxygen content of liquid steel is often required. It is done by oxygen sensor, which is an emf cell with SE electrolyte. The sensor is immersed into liquid steel. Since it gets damaged, it cannot be reused. The sensor also carries a thermocouple. The readings of emf and temperature are to be completed within a minute.

Several commercial designs are available. A common design is:

$$\text{Mo} \mid \text{Cr} + \text{Cr}_2\text{O}_3 \mid \text{ZrO}_2\text{–MgO} \mid \text{liquid steel} \mid \text{Fe} \qquad \text{(E.13.24)}$$
$$\text{(reference)}$$

Molybdenum and iron serve as leads. Since they are dissimilar, emf reading requires *correction of the thermocouple effect*. The solid electrolyte is in the form of a small tube open at one end.

From Eq. (E.13.16),

$$\mu_{O_2} \text{ (liquid steel)} - \mu_{O_2} \text{ (reference)} = -4FE \qquad \text{(E.13.25)}$$

Since Cr and Cr_2O_3 are pure, we obtain the following relation on the basis of Eq.(8.10):

$$\mu_{O_2} \text{ (reference)} = RT \ln p_{O_2} \text{ (reference)}$$
$$= \Delta G_f^\circ \text{ for formation of } 2/3\text{Cr}_2\text{O}_3(s) \text{ (i.e. per mol.O}_2) \qquad \text{(E.13.26)}$$

For liquid steel, the relevant reaction is

$$\frac{1}{2}O_2(g) = [O]_{\text{dissolved in liquid steel}} \qquad \text{(E.13.27)}$$

for which the equilibrium relation is

$$K_{27} = \left\{ \frac{[h_O]}{p_{O_2}^{1/2}} \right\}_{eq} \qquad \text{(E.13.28)}$$

where K_{27} is equilibrium constant for reaction (E.13.27) and h_O is the activity of dissolved oxygen in steel in 1 wt.% standard state (see Section 11.5). Hence,

$$\mu_{O_2} \text{ (liquid steel)} = RT \ln p_{O_2} \text{ (in equilibrium with liquid steel)}$$

$$= 2RT \ln \left[\frac{h_O}{K_{27}} \right] \qquad \text{(E.13.29)}$$

Combining these with Eq. (E.13.25), we get

$$2RT \ln \left[\frac{h_O}{K_{27}} \right] - \Delta G_f^0 \left(\frac{2}{3} Cr_2O_3 \right) = -4FE \qquad (E.13.30)$$

At 1873 K, ΔG_f^0 of $Cr_2O_3 = -422.7 \times 10^3$ J/mol O_2, $K_{27} = 2615$. If $E = -0.153$ V, then from Eq. (E.13.30), $h_O = 0.0005$. If the behaviour of oxygen in steel is assumed to be Henrian, then

$$\text{weight \% O} = W_O = h_O = 0.0005 \text{ wt.\% (i.e. 5 ppm)} \quad [\text{Ans.}] \qquad (E.13.31)$$

13.4 Single Electrode Potentials and Electrochemical Series in Aqueous Systems

The electrical potential of a single electrode is unknown. It cannot be found out since any measurement would require two electrodes. Therefore, the term *single electrode potential ε,* as used in literature, is not literally correct from the above point of view, and requires clarification.

H^+ is the common ion in aqueous solutions. Hence, the single electrode potential of a standard hydrogen electrode has been arbitrarily assumed to be zero, and is employed as *reference electrode* for measurements of ε of other electrodes. For example, let us consider the following cell set-up:

$$\text{Zn(s)} \mid \text{ZnCl}_2\text{(aq)} \parallel \text{HCl(aq)} \mid \text{Pt(s), H}_2\text{(g)} \qquad (13.22)$$
$$(a_{Zn^{2+}} = 1) \qquad (a_{H^+} = 1) \quad \text{at 1 atm}$$

The emf of this cell is taken as the standard single electrode potential of Zn/Zn²⁺ couple. It is designated as $\varepsilon^0_{Zn/Zn^{2+}}$. Again, platinum serves as the hydrogen electrode since its surface comes to equilibrium with the gas fairly fast.

There are two conventions for single electrode potential, viz. oxidation potential and reduction potential.

(i) *Oxidation potential.* For oxidation reaction such as

$$\text{Zn(s)} = \text{Zn}^{2+}\text{(aq)} + 2e^-, \qquad \varepsilon^0 = +0.763 \text{ V} \qquad (13.23)$$

(ii) *Reduction potential.* For reduction reaction such as

$$\text{Zn}^{2+}\text{(aq)} + 2e^- = \text{Zn(s)}, \qquad \varepsilon^0 = -0.763 \text{ V} \qquad (13.24)$$

In other words, reduction potential of a system is equal and opposite to that of its oxidation potential.

In conformity with other standard text books, Table 13.2 presents values of *standard reduction potentials* for some metals at 25°C in aqueous solutions of their respective ions. The standard single electrode potential (ε^0) is the value of ε, when both metal and its ion are at their respective standard states, as in cell (13.22), where $a_{Zn}, a_{Zn^{2+}}, a_{H^+}, p_{H_2}$ are all equal to 1. Here, the question arises–what do we mean by $a_{Zn^{2+}} = 1$ in an aqueous solution? $a_{Zn^{2+}}$ simply can never by measured. Hence, by convention, if the concentration of the salt ($ZnCl_2$) is 1 molal (i.e. 1 gm-mole in 1000 g water), then ε is taken as ε^0. This is indicated in Table 13.2. If the ion exhibits Henrian behaviour, i.e. activity is proportional to molality, then the calculation of ε at concentrations other than 1 molal is simple. However, it is not true in most cases and, hence, specific information of behaviour is required. Further discussion of this topic is beyond the scope of this text.

Table 13.2   Standard single electrode potentials in aqueous solution at 298 K, 1 atm (standard state is 1 molal)

Electrode reaction	ε^0, volts
$F_2 + 2e^- = 2F^-$	2.65
$Co^{3-} + e^- = Co^{2-}$	1.82
$Cl_2 + 2e^- = 2Cl^-$	1.3595
$O_3 + H_2O + 2e^- = O_2 + 2OH^-$	1.24
$Br_2(l) + 2e^- = 2Br^-$	1.0652
$2Hg^{2+} + 2e^- = Hg_2^{2+}$	0.92
$Hg^{2+} + 2e^- = Hg$	0.854
$Ag^+ + e^- = Ag$	0.7991
$Fe^{3+} + e^- = Fe^{2+}$	0.771
$I_2 + 2e^- = 2I^-$	0.5355
$Cu^{2+} + 2e^- = Cu$	0.337
$Cu^{2+} + e^- = Cu^+$	0.153
$2H^+ + 2e^- = H_2$	0.000
$Fe^{3+} + 3e^- = Fe$	-0.036
$Pb^{2+} + 2e^- = Pb$	-0.126
$Sn^{2+} + 2e^- = Sn$	-0.136
$Cd^{2+} + 2e^- = Cd$	-0.403
$Cr^{3+} + e^- = Cr^{2+}$	-0.41
$Fe^{2+} + 2e^- = Fe$	-0.440
$Zn^{2+} + 2e^- = Zn$	-0.763
$2H_2O + 2e^- = H_2 + 2OH^-$	-0.828
$ZnO_2^{2-} + 2H_2O + 2e^- = Zn + 4OH^-$	-1.216
$Al^{3+} + 3e^- = Al$	-1.66
$Mg^{2+} + 2e^- = Mg$	-2.37
$Na^+ + e^- = Na$	-2.714
$Ca^{2+} + 2e^- = Ca$	-2.87
$K^+ + e^- = K$	-2.925
$Li^+ + e^- = Li$	-3.045

The overall reaction in a Daniel cell is given by Eq. (13.2). It is the sum of Eqs. (13.3) and (13.4). Hence, for the Daniel cell,

$$\text{Standard cell emf } (E^0) = \mathcal{E}^0_{Zn/Zn^{2+}} + \mathcal{E}^0_{Cu^{2+}/Cu} = \mathcal{E}^0_{Cu^{2+}/Cu} - \mathcal{E}^0_{Zn^{2+}/Zn}$$

$$= 0.337 - (-0.763) = 1.10 \text{ V at } 25°C \qquad (13.25)$$

If the electrode and electrolyte are not at their standard states then, on the basis of Eq. (13.11), the reduction potential for Zn^{2+}/Zn is given as

$$\varepsilon = \mathcal{E}^0 - \frac{RT}{2F} \ln \left[\frac{a_{Zn}}{a_{Zn^{2+}}} \right] \qquad (13.26)$$

In Eq. (13.26), Zn is the reduced state and Zn^{2+} is the oxidized state. We may generalize it as

$$\text{Oxidized state} + 2e^- = \text{reduced state} \qquad (13.27)$$

On the basis of the above equations, the generalized equation is

$$\varepsilon = \mathcal{E}^0 - \frac{RT}{ZF} \ln \left(\frac{a_R}{a_O} \right) \qquad (13.28)$$

where a_O = activity of oxidized species and a_R = activity of reduced species. Equation (13.28) is the famous *Nernst equation* in electrochemistry.

13.5 Pourbaix Diagrams

Pourbaix diagrams, or *potential pH diagrams,* are graphical representations of electrochemical equilibria occurring in aqueous systems, where pH = $-\log$ [H^+]. It was first proposed by Pourbaix* and are very useful in the areas of corrosion, electrometallurgy and hydrometallurgy. They are essentially electrochemical analogues of phase stability diagrams discussed in Chapters 10 and 11.

As an example, again consider dissolution of Zn in aqueous solution. At 25°C, in acidic solution of pH < 5, the metal dissolves in aqueous solution as Zn^{2+}. At higher pH, $Zn(OH)_2$ is the thermodynamically stable phase, and the equilibrium activity of Zn^{2+} will be governed by the solubility product of $Zn(OH)_2$ and the pH of the solution, i.e. by the equilibrium of the reaction

$$2H^+ + Zn(OH)_2(s) = Zn^{2+} + 2H_2O \qquad (13.29)$$

At still higher pH, ZnO_2^- ions become stable and both Zn^{2+} and $Zn(OH)_2$ are unstable.

---

*M. Pourbaix, *J. Electrochem, Soc.*, **101**, p. 213, 1954.

The above discussion is applicable to a situation where there is no externally imposed electric potential on the system. By imposing a potential, however, pH ranges for stability of phases can be altered. Here, the *guiding criterion is the Nernst equation*. For a Zn^{2+}/Zn couple, it is Eq. (13.26). If the activity of Zn^{2+} in aqueous solution and temperature are known, ε for Zn^{2+}/Zn can be calculated from Eq. (13.26). If the externally imposed potential (ε_{ext}) is more negative than ε, then metallic zinc is stable. If ε_{ext} is less negative than ε, then Zn^{2+} is stable. Also, it should be noted that the activities in various phases would govern stabilities.

Figure 13.2 gives the potential-pH diagram for Zn/H_2O system at 25°C. The stable phases are indicated. The reactions and equations for construction of Fig. 13.2 are as follows:

$$Zn^{2+} + 2e^- = Zn, \qquad \varepsilon = \varepsilon^0 + 0.03 \log (a_{Zn^{2+}}) \qquad (13.30)$$

$$Zn(OH)_2 + 2H^+ + 2e^- = Zn + 2H_2O, \qquad \varepsilon = \varepsilon^0 - 0.059 \text{ pH} \qquad (13.31)$$

$$ZnO_2^{2-} + 4H^+ + 2e^- = Zn(OH)_2 + H_2, \qquad \varepsilon = \varepsilon^0 - 0.118 \text{ pH} + 0.03 \log (a_{ZnO^{2-}}) \qquad (13.32)$$

where "log" is the common logarithm.

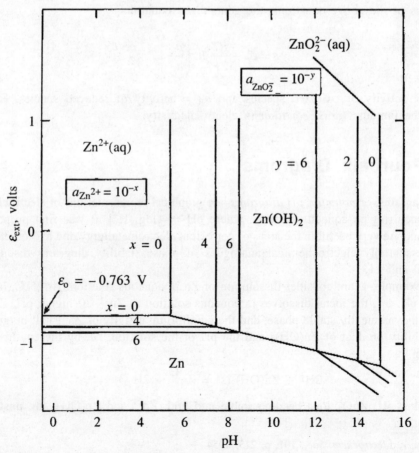

Fig. 13.2 Simplified potential-pH diagram for Zn/H_2O system at 298 K.

From Table 13.2, $\varepsilon_0 = -0.763$. For construction of Fig. 13.2, three values of a_{Zn}^{2+} and $a_{ZnO^{2-}}$ were assumed as 1, 10^{-4} and 10^{-6} and 1, 10^{-2} and 10^{-6} respectively.

13.6 Summary

1. Electrochemical cells may be classified into galvanic cells (i.e. batteries) and electrolytic cells. Only galvanic cells can be operated reversibly.

2. Besides aqueous solutions at room temperature, molten salts, molten slags (i.e. oxides) and solid electrolytes are employed as electrolytes in metallurgy and materials applications at high temperatures.

3. For a reversible galvanic cell, the cell emf(E) is related to thermodynamic quantities by the two alternate forms of the equation:

   (i) $$\Delta G = -ZFE$$

   where ΔG is free energy of cell reaction, Z is valency involved, and F is Faraday's constant.

   (ii) $$\mu_i^{II} - \mu_i^{I} = -ZFE$$

   where i is the species exchanged between electrodes I and II, and μ_i is the chemical potential of i. The first form is used for reaction cells and the second form is preferred for concentration cells.

4. $$E = E^0 - \frac{RT}{ZF} \ln J$$

   where E^0 is standard emf of the cell, and J is the activity quotient.

5. Some examples of emf cells have been discussed in this chapter. It may be specially mentioned that, for a solid oxide electrolyte cell, equation in 3(ii) above may be rewritten as

   $$E = \frac{RT}{4F} \ln \frac{p_{O_2}^{I}}{p_{O_2}^{II}}$$

   where $p_{O_2}^{I}$ and $p_{O_2}^{II}$ are partial pressures of oxygen at electrodes I and II, respectively. The US Sign convention for emf has been adopted in this text.

6. The single electrode potential of an electrode (ε), and electrochemical series in aqueous solution have been discussed. ε can be taken as oxidation potential or reduction potential, which are equal and opposite for a specific system. The reduction potential has been chosen for application in this text.

7. The fundamentals of Pourbaix diagrams or Potential-pH diagrams, have been illustrated through the example of Zn-aqueous solution system.

PROBLEMS

13.1 The emf of the cell

$$\text{Pt, } H_2(g) \mid H_2SO_4 \mid PbO_2(s) + PbSO_4(s), \text{ Pt}$$
$$\text{aqueous}$$
$$\text{solution}$$

was measured as 1.63195 and 1.62950 volts, respectively at 45°C and 35°C. Calculate ΔG, ΔH and ΔS of the cell reaction at 25°C.

13.2 The emf of the cell: $Ag(s) \mid AgCl(s) \mid Cl_2(1 \text{ atm})$, Pt, was found to be:

$$E(\text{volts}) = 0.977 + 5.7 \times 10^{-4} (350 - t) - 4.8 \times 10^{-7} (350 - t)^2$$

in the temperature range $t = 100°C$ to $450°C$. Calculate the value of ΔC_P for the cell reaction.

13.3 For commercial production of magnesium, $MgCl_2$ is decomposed by electrolysis according to the following reaction at high temperature:

$$MgCl_2(l) = Mg(l) + Cl_2(g), \text{ 1 atm}$$

Calculate the minimum theoretical voltage (to overcome cell emf) to be applied for decomposition at 750°C. Assume $MgCl_2$, Mg as pure.

13.4 The emf of the cell: $Al(s) \mid AlCl_3–NaCl$ melt $\mid$ Al-Zn alloy(s)

was measured as 7.43 millivolt at 380°C, and the temperature coefficient of the emf as 2.9×10^{-5} volt/°C. Calculate a_{Al}, $\bar{G}_{Al}^m$ and $\bar{H}_{Al}^m$ in the alloy.

13.5 Consider the high temperature emf cell in Eq. (E.13.18).

At $X_{Sn} = 0.5$ in the Sn-Ag alloy, a measured cell emf was fitted with the equation: $E(\text{volt}) = 14.69 \times 10^{-3} - 30.4 \times 10^{-6} T$, where T is temperature in K. Calculate (at $X_{Sn} = 0.5$):

(i) $\qquad\qquad\qquad a_{Sn}, \ \bar{G}_{Sn}^m, \bar{G}_{Sn}^{XS}$ at 1200 K

(ii) $\qquad\qquad\qquad \bar{S}_{Sn}^m, \bar{S}_{Sn}^{XS}, \bar{H}_{Sn}^m, \bar{H}_{Sn}^{XS}$

13.6 Calculate the emf of the following electrochemical cell at 1200 K:

$$\text{Pt} \mid Ni(s) + NiO(s) \mid ZrO_2 - CaO \mid Mn(s) + MnO(s) \mid \text{Pt}$$

13.7 Consider the following solid electrolyte cell:

$$\text{Pt} \mid SO_2 + SO_3 \text{ gas mixture} \mid ZrO_2 - CaO \mid M(s) + MO(s) \mid \text{Pt}$$

Calculate:

(i) ΔG^0 of formation of MO

(ii) chemical potential of S_2 in the gas mixutre.

Given: $T = 1500$ K; total pressure of $SO_2 + SO_3 = 1$ atm;

p_{SO_2}/p_{SO_3} ratio = 500; cell emf = 0.46 volt

13.8 The emf of the following cell:

$$Cu(s) + Cu_2O(s) \mid ZrO_2 - CaO \mid Cu - Ag \text{ alloy (l)} + Cu_2O(s)$$

was measured as: $E(mV) = -177.8 + 0.1657\, T$,

at $X_{Cu} = 0.5$ and temperature range 1090–1300 K.

Calculate (at 1200 K):

 (i) a_{Cu} with solid Cu as standard state

 (ii) a_{Cu} with liquid Cu as standard state.

(If the sign of emf causes confusion, use your judgement. For example, the calculated activity of a component in alloy should be less than 1.)

Chapter 14

Thermodynamics of Surfaces, Interfaces and Defects

14.1 Introduction

By *surface*, we mean a free surface which separates a solid or liquid from surrounding gas or vacuum. Ideally, the surrounding should be a vacuum. However, a gas phase has negligible concentration of atoms/molecules per unit volume as compared to that in solid or liquid. Hence, approximately any interface with gas is also treated as a free surface.

Surfaces play an important role in a variety of phenomena encountered in metallurgy and materials science. A surface is only a few atomic layers thick. On the surface, there are some free bonds projecting normal to surface. This causes a difference in structure of the surface layer from that of the bulk of the solid/liquid.

A surface possesses some extra energy due to the work done to create it. Specific surface energy is this excess energy per unit surface area (say, in joules per m^2 of surface). From an atomistic viewpoint, work is to be done to create free bonds on the surface (i.e. equivalent to bond breaking). Bond energy is of the nature of enthalpy. Strictly speaking, it is internal energy, which is approximately the same as enthalpy for condensed phases. Since $H = G + TS$, excess surface energy is due to excess surface free energy and excess surface entropy. But specific surface energy is usually taken as free energy (either Gibbs or Helmholtz free energy, depending on application) per unit surface area, in excess of that of the bulk.

An *interface,* by convention, is the *surface separating two condensed phases.* Here, we do not have the concept of free bonds. Solid surfaces are rough. Hence, the interface between two solids has only some contact points. Moreover, such contacts depend on surface roughness, and hence, are unpredictable and irreproducible. Therefore, thermodynamics is not capable of predicting its properties. Polished interfaces joined by bonding such as adhesives, soldering, brazing etc., are more amenable to thermodynamic treatment. In contrast, solid/liquid and liquid/liquid interfaces, especially the latter, can be more definitely reproducible. Excess energy at the interface is influenced by the extent of composition mismatch and/or structural mismatch (such as water and ice). At electrode/electrolyte, slag/metal, metal/aqueous solution interfaces, we have *the electric double layer* also, which modifies interfacial energy. Hence, the phenomena here are more complex than those at free surfaces. Grain boundaries in solids are also interfaces. There, the mismatch in grain orientations gives rise to excess interfacial energy (i.e. *grain boundary energy)*.

Grain boundaries, stacking faults, point defects (i.e. vacancies and interstitials) and line defects (i.e. dislocations) play very important roles in phase transformations, diffusion etc. and in control of mechanical and other properties of solids, as well as their processing. Fine particles have very high surface energy per unit volume, which are responsible for their special behaviour pattern.

In all the above areas, it is thermodynamics again, which constitutes a principal scientific foundation. However, other courses in the area of metallurgy and materials are offered, especially on structure, properties and processing of solid materials, where thermodynamic aspects are also dealt with in detail. Hence, for the sake of basic information and understanding, in this introductory text, only a brief mention will be made about thermodynamic aspects.

14.2 Thermodynamics of Surfaces

14.2.1 General Considerations

From Chapter 10, we recall that, at constant temperature and pressure,

$$dG' = \mu_1 \, dn_1 + \mu_2 \, dn_2 + \ldots + \mu_i \, dn_i + \ldots = \sum_i \mu_i \, dn_i \qquad (10.11)$$

where 1, 2, ... are components, in the absence of any other work. In the presence of any other work ($\delta W'$),

$$dG' = \sum_i \mu_i \, dn_i - \delta W' \qquad (14.1)$$

The work done to create an infinitesimal small surface dA is given as

$$\delta W' = -\sigma \, dA \qquad (14.2)$$

where the negative sign is due to the fact that work is to be done on the surface by an outside agency to create a surface; σ *is specific surface energy*, as defined in Section 14.1. Therefore,

$$dG' = \sum_i \mu_i \, dn_i + \sigma \, dA \qquad (14.3)$$

$\delta W'$ is nothing but $[-dG'_s]$, which is the *change in surface free energy*. Hence,

$$dG'_s = \sigma \, dA \qquad (14.4)$$

If σ is independent of surface area, then integration of Eq. (14.4) leads to

$$G'_s = \sigma A, \quad \text{i.e.} \quad \sigma = \frac{G'_s}{A} = G_{SA} \tag{14.5}$$

where G'_s is surface free energy and G_{SA} is specific surface energy per unit area. σ is a constant for a surface, provided the surface is *isotropic*, i.e. it has no directionality. Liquid surfaces are isotropic, whereas solid surfaces may or may not be isotropic.

You are familiar with *surface tension*, which is force per unit length of a surface. This originated from Young's model, which considered the mechanical behaviour of a surface. The actual interface is replaced by an imaginary infinitesimally thin elastic membrane, called surface of tension. Dimensions of specific surface energy and *surface tension* are the same (Mt^{-2}). *For isotropic surfaces, numerically also they are equal.* Table 14.1 presents some values of σ.

Table 14.1 Specific Surface Energies of Some Materials

Material	Temperature, K	$G_{SA} \times 10^3$, J m^{-2}
Water (l)	298	72
NaCl (s)	298	300
Al$_2$O$_3$ (s)	2123	905
Al$_2$O$_3$ (l)	2353	700
TiC (s)	1373	1190
Al (s)	723	980
Al (l)	933	866
Cu (s)	1198	1780
Cu (l)	1573	1300
Fe (l)	1823	1835
Pb (l)	600	450
Hg (l)	298	465

14.2.2 Adsorption Isotherms

Surface free energy is lowered by the phenomenon of adsorption, and this lowering constitutes the driving force for the same. Adsorption means preferential incorporation of a component i (element or compound) in the surface layer. In other words, there is an excess concentration of i on the surface as compared to that in the bulk, if component i is adsorbed on the surface.

Adsorption is broadly classified into physical adsorption and chemisorption.

(i) *Physical adsorption.* Here, the forces between the surface and the adsorbed species is physical in nature (van der Waals type forces)

(ii) *Chemisorption.* Where, chemical bonds form between surface and adsorbed species.

Physical adsorption is dominant at low temperatures. Above room temperature, chemisorption is dominant. Hence we are concerned primarily with the latter, and subsequent discussions will be restricted to only chemisorption.

Adsorption isotherm means adsorption equilibrium relations at constant temperature. For chemisorption, we are concerned with the following adsorption isotherms:

    (i) Langmuir adsorption isotherm
    (ii) Gibbs adsorption isotherm

These are according to the names of the scientists who proposed them.

Langmuir adsorption isotherm

This was proposed by Langmuir for adsorption of gases on solid/liquid surfaces. The assumptions are:

1. The adsorbed molecules/atoms are present on the surface as a single layer (i.e. *monolayer adsorption*)
2. The gas is ideal
3. Adsorbed layer is ideal, i.e. there is no interaction amongst adsorbed atoms/molecules.

The process of chemisorption was considered as a chemical reaction by Langmuir, and was written in the form

$$G + \overset{|}{S} = \overset{\overset{G}{|}}{\underset{|}{S}} \tag{14.6}$$

where G is the symbol for a monatomic gas molecule, $\overset{|}{S}$ indicates a free surface site, $\overset{\overset{G}{|}}{\underset{|}{S}}$ is a surface site with adsorbed gas atom.

If K_a is adsorption equilibrium constant at a temperature, then

$$\frac{\theta}{1 - \theta} = K_a p_G \tag{14.7}$$

where
    $1 - \theta$ = fraction of free surface
    θ = fraction of surface site with adsorbed atom.

θ, $(1-\theta)$ are measures of surface concentration, i.e. mole fractions of surface sites. Since the atoms in the adsorbed layer behave ideally, mole fractions are the same as respective activities. These considerations constitute the basis of Eq. (14.7), which is the *Langmuir adsorption isotherm*.

For adsorption of diatomic gases, adsorption is preceded by dissociation of molecules, i.e.

$$\frac{1}{2} G_2(g) = G(g), \quad K_8 = \frac{p_G}{p_{G_2}^{1/2}} \tag{14.8}$$

From Eqs. (14.7) and (14.8),

$$\frac{\theta}{1 - \theta} = K_a K_8 \, p_{G_2}^{1/2} \tag{14.9}$$

It may be noted that Langmuir isotherm can also be applied to adsorption of a component from the bulk of the solution. This requires invoking some additional equilibrium relations.

Gibbs adsorption isotherm

The Gibbs adsorption isotherm is applicable to adsorption of a component i (element/compound) of a solution onto the surface of the solution. Diffusion of atoms and molecules are several orders of magnitude faster in liquid as compared to those in solids. Moreover, convection in a liquid enhances mass transfer significantly. Hence, attainment of adsorption equilibrium between the surface and the bulk of a solution is possible only for liquid solutions in a realistic timeframe.

For Gibbs free energy, Eq. (9.16) may be written as

$$G' = \sum_i \bar{G}_i n_i \tag{14.10}$$

Since we are considering a surface, the surface energy term is to be included in Eq. (14.10). Hence,

$$G' = \sum_i \bar{G}_i n_i + A\sigma \tag{14.11}$$

Differentiating Eq. (14.11), we get

$$dG' = \left(\sum_i \bar{G}_i \, dn_i + \sigma \, dA \right) + \left(\sum_i n_i \, d\bar{G}_i + A \, d\sigma \right) \tag{14.12}$$

Replacing $\bar{G}_i$ by μ_i, we get

$$dG' = \left(\sum_i \mu_i \, dn_i + \sigma \, dA \right) + \left(\sum_i n_i \, d\mu_i + A \, d\sigma \right) \tag{14.13}$$

Combining Eqs. (14.13) and (14.3), we obtain the *modified Gibbs-Duhem equation for surfaces* as

$$\sum_i n_i \, d\mu_i + A \, d\sigma = 0 \tag{14.14}$$

Consider adsorption of component i only. Then,

$$n_i \, d\mu_i + A \, d\sigma = 0 \tag{14.15}$$

i.e.

$$\frac{n_i}{A} = -\frac{d\sigma}{d\mu_i} \qquad (14.16)$$

By definition,

$$\text{Surface excess of } i = \Gamma_i = \left(\frac{n_i}{A}\right)_{\text{surface}} - \left(\frac{n_i}{A}\right)_{\text{bulk}} \qquad (14.17)$$

Since, for a surface active species,

$$\left(\frac{n_i}{A}\right)_{\text{surface}} \gg \left(\frac{n_i}{A}\right)_{\text{bulk}} \qquad (14.18)$$

$$\Gamma_i \approx \left(\frac{n_i}{A}\right)_{\text{surface}} = -\frac{d\sigma}{d\mu_i} \qquad (14.19)$$

Equation (14.19) is the *mathematical statement of Gibbs adsorption isotherm*. One of the best examples of its application is for adsorption of oxygen atoms on surface of liquid iron at high temperature, if we dissolve some oxygen in the liquid.

Figure 14.1 shows variation of σ with wt.% oxygen dissolved in liquid iron, as measured experimentally. Since $d\sigma/d\mu_i$ is negative in this case, Γ_i is positive from Eq. (14.19). Thus, oxygen is preferentially adsorbed on surface of liquid iron, i.e. it is *surface active*.

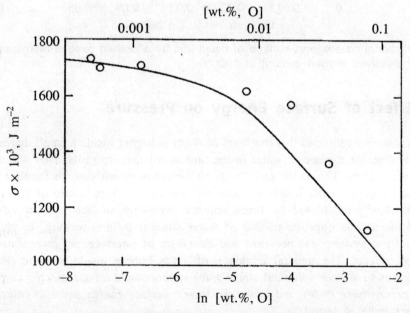

Fig. 14.1 Variation of surface tension of liquid iron with dissolved oxygen content at 1550°C [F.A. Holden and W.D. Kingery, *J. Phys. Chem.*, **59**, 557, 1955].

Figure 14.2 shows fraction of surface covered by adsorbed oxygen as function of wt.% oxygen in the iron-oxygen solution. It was calculated from the data given in Fig. 14.1 assuming monolayer adsorption. It may be noted that the surface is almost saturated with oxygen even at as low as 0.03% oxygen. In other words, oxygen is strongly surface active in liquid iron. Sulphur also shows a similar behaviour. It may be mentioned that these phenomena significantly influence steel making processes.

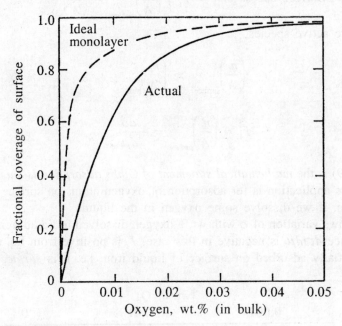

Fig. 14.2 Fractional coverage of surface of liquid iron by adsorbed oxygen atoms as function of dissolved oxygen content at 1600°C.

14.2.3 Effect of Surface Energy on Pressure

It had been known for centuries that the level of water is higher inside a small diameter glass tube. Not only that, the surface of water in the tube is concave upwards. This was attributed to *water 'wetting' glass*. This is the *capillarity phenomenon* which you are familiar with. All these, viz. the height of water inside the tube and the radius of curvature of water surface, can be quantitatively calculated by force balance involving surface tension values. The behaviour of mercury is opposite to that of water since it is 'non-wetting' to glass.

In a liquid-gas system, gas pressure and curvature of interface are interrelated due to surface energy effects. The general tendency of a gas bubble inside a liquid or a liquid droplet in a gas is to assume spherical shape *since spheres have lowest specific surface area (i.e. surface area/volume ratio)*, and hence the lowest surface energy per unit volume of the liquid (i.e. per mole of liquid).

Excess pressure (ΔP) inside a gas bubble in liquid

Consider a gas bubble of radius r inside a liquid. Let ΔG_{XS} be the excess free energy due to the presence of the bubble. Then,

$$\Delta G_{XS} = 4\pi r^2 \sigma - \frac{4}{3}\pi r^3 \left(\frac{\Delta G_p}{V}\right) \tag{14.20}$$

where $\Delta G_p/V$ is the excess free energy of bubble per unit volume due to excess pressure (ΔP) inside the bubble. The excess pressure is due to squeezing action of surface tension force on the bubble. Hence, the 2nd term in Eq. (14.20) containing ΔG_p is negative. At constant temperature, $dG = VdP$. Hence,

$$\frac{\Delta G_p}{V} = \Delta P \tag{14.21}$$

Combining Eqs. (14.20) and (14.21), we get

$$\Delta G_{XS} = 4\pi r^2 \sigma - 4/3\pi r^3 \cdot \Delta P \tag{14.22}$$

The system would have the minimum ΔG_{XS} at bubble-liquid equilibrium. Therefore, at equilibrium,

$$\frac{d(\Delta G_{XS})}{dr} = 0 = 8\pi r \sigma - 4\pi r^2 \cdot \Delta P \tag{14.23}$$

i.e.

$$\Delta P = \frac{2\sigma}{r} \tag{14.24}$$

The value of ΔP can be quite large for tiny bubbles. For example, if the bubble radius is 0.1 mm (10^{-4}m), then $\Delta P = 3.2 \times 10^4$ Pa (0.3 atm) in liquid steel with $\sigma = 1.6$ J m^{-2}, at bubble-steel equilibrium.

Equilibrium vapour pressure of a liquid droplet

Consider a pure liquid A. For a large volume with flat surface, its saturated vapour pressure (p_A^0) is given by the Clausius–Clapeyron equation (section 6.4.2). Proceeding with derivation along the line as for gas bubble in liquid, it can be shown that the equilibrium vapour pressure of the droplet (p_A) will be larger than p_A^0, and would follow the relation

$$p_A = p_A^0 + \frac{2\sigma}{r} \tag{14.25}$$

where r is the droplet radius.

Figure 14.3 shows variation of vapour pressure of liquid zinc as function of droplet radius at 900 K.

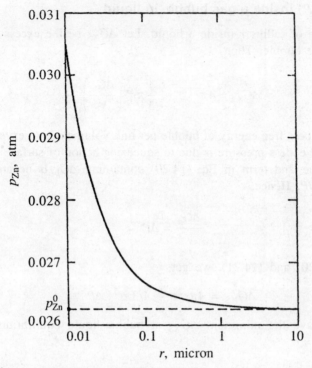

Fig. 14.3 Equilibrium vapour pressure of zinc at 900 K as function of radius of liquid Zn droplet.

14.2.4 Effect of Surface/Interface Energy on Equilibrium Phase Transformation Temperature

In Section 6.4, we discussed equilibrium phase transformation in condensed systems. Figure 6.1 provided basic classifications by a schematic presentation of $G_A(I)$, $G_A(II)$ and $\Delta G[=G_A(II) - G_A(I)]$ at constant pressure as function of temperature. In the figure, phase II is stabler than phase I above the equilibrium transformation temperature, T_{tr}.

Let us now suppose that A is present in phase I in fine sizes with a large specific surface area. Then $G_A(I)$ will change to $G_A^*(I)$, where

$$G_A^*(I) = G_A(I) + G_S \qquad (14.26)$$

with G_S as surface energy per mole of A in phase I. Then,

$$\Delta G^* = G_A(II) - [G_A(I) + G_S] = \Delta G - G_S \qquad (14.27)$$

Figure 14.4 schematically presents the features of Eq. (14.27). As may be noted from the figure, T_{tr} gets lowered to T_{tr}^* due to surface energy of phase I. This phenomenon is encountered during melting of fine particles of solid, or crystals with a highly curved surface, such as at tips of dendrites, where melting occurs at a lower temperature. It is obvious that the situation will be reverse and T_{tr} will increase if phase II is of fine size.

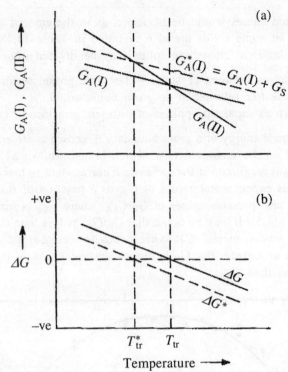

Fig. 14.4   Illustration of effect of surface energy on equilibrium phase transformation temperature at constant pressure (schematic).

14.3   Thermodynamics of Defects in Solids

As stated in Section 14.1, defects in solids govern a variety of properties of solids—mechanical, electrical, electronic, and so on. Hence, they are elaborately taught in metallurgy and materials science programme. Thermodynamics of defects are also covered there. Therefore, we shall make only very brief mention of some important points here to understand the basics.

It should be mentioned here that concentration of dislocations is orders of magnitude larger than what thermodynamic equilibrium predicts. It is a highly nonequilibrium situation and thermodynamics is not of much help. Hence, we shall not discuss about dislocations.

14.3.1   Surfaces and Interfaces of Solids

Liquid surfaces are isotropic. Also, atoms at liquid surfaces are mobile, and hence, equilibrium is attained easily. But these are not generally applicable to solid surfaces. Therefore, the meaning of surface tension is not clearcut for solids. Also, specific surface energy and surface tension would not have the same numerical value. Consider a single crystal. It does not possess a unique specific surface energy since the latter has different values for different crystal planes. However, for a polycrystalline solid with random orientations at the surface, the statistically averaged values would be isotropic, as in the case of a liquid surface. The surface energy values of some solids are also included in Table 14.1.

The internal boundaries in crystalline solids may be divided into the following:

1. Internal boundaries between crystals of the same phase, resulting from orientation differences (hereafter referred to as *grain boundaries*)
2. Internal boundaries separating phases of different structure or compositions or both.

The specific interfacial energy of a grain boundary is known as *its grain boundary energy* G_B. G_B would depend on the misorientation, which is also known as the *tilt angle (θ)* of neighbouring grains. G_B is maximum at $\theta = 45°$ since it corresponds to maximum misorientation.

Figure 14.5 presents experimental values of G_B, as a function of θ in copper at 1338 K. Besides G_B, the figure also presents values of G_B/G_{SA}, where G_{SA} is surface energy per unit area, as defined in Eq. (14.5). It may be noted that G_B/G_{SA} is less than 0.4. It has been found to be less than 0.5 for several metals. G_{SA} corresponds to increase of free energy due to the presence of free bonds at surface. But G_B is due to misorientation of bonds only. Thus, G_B is expected to be lower than G_{SA}.

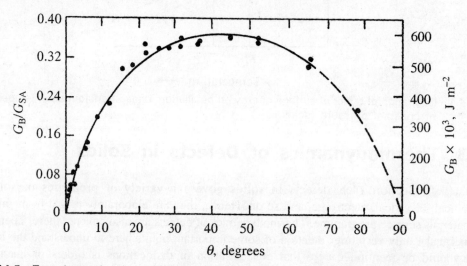

Fig. 14.5 Experimental value of grain boundary energy as function of misorientation for a simple tilt boundary in copper at 1338 K [N.A. Gjostein and F.N. Rhines, *Acta Met.*, 7, 319, 1959].

14.3.2 Vacancies and Interstitials in Solid Metals

Derivation of equilibrium relations

Vacancies and interstitials are point defects. As an illustration, let us consider vacancies in a crystalline metal. In N atoms of a perfect crystal, the number of atoms as well as the number of lattice sites are also N. If n_v number of atoms are removed from lattice points and transferred to the surface, we create n_v vacancies. Hence, mole fraction of vacancies (X_v) would be equal to $n_v/(N+n_v)$, and mole fraction of atoms = $N/(N+n_v) = 1 - X_v$.

Now, $N + n_v$ corresponds to $(N+n_v)/N_0 = y$ moles, where N_0 is Avogadro's number. Hence, the number of moles of vacancies $= yX_v$. Let,

ΔH_v = Enthalpy of formation of 1 mole of vacancies
ΔS_v = Change in vibrational entropy due to vacancy formation, per mole of vacancy
ΔS_c = Configurational entropy change due to mixing of n_v vacancies in lattice sites.
Then, from Chapter 12,

$$\Delta S_c = k_B \ln \frac{W'}{W} \tag{14.28}$$

For a perfect crystal, the number of arrangements, $W = 1$. W' is the number of arrangements of n_v vacancies in $N + n_v$ lattice sites. From Eq. (12.23) and Stirling's approximation [Eq.(12.24)],

$$\Delta S_c = k_B \ln W' = k_B \ln \frac{(N + n_v)!}{N! \, n_v!} = -k_B \left[N \ln \frac{N}{N + n_v} + n_v \ln \frac{n_v}{N + n_v} \right] \tag{14.29}$$

In terms of mole fractions,

$$\Delta S_c = -k_B[N \ln(1 - X_v) + n_v \ln X_v] \tag{14.30}$$

If ΔG is the free energy change due to formation of n_v number of vacancies, then

$$\Delta G = yX_v(\Delta H_v - T\Delta S_v) - T \Delta S_c \tag{14.31}$$

Combining Eq. (14.31) with Eq. (14.29), we get

$$\Delta G = \frac{n_v}{N_0} (\Delta H_v - T\Delta S_v) + \frac{RT}{N_0} \left(N \ln \frac{N}{N + n_v} + n_v \ln \frac{n_v}{N + n_v} \right) \tag{14.32}$$

Figure 14.6 presents the various terms of Eq. (14.32) schematically as function of vacancy concentration n_v. R and N_0 are universal constants. T, N, ΔH_v, ΔS_v are also assumed constant in Fig. 14.6.

It may be noted that ΔG goes through a minimum. The equilibrium vacancy concentration, $n_v(\text{eq})$, corresponds to this minimum. At this concentration, $d(\Delta G)/dn_v = 0$.
Differentiation of Eq. (14.32) leads to

$$N_0 \cdot \frac{d(\Delta G)}{dn_v} = \Delta H_v - T\Delta S_v - RT \left[-\frac{N}{N + n_v} + (\ln n_v - \ln(N + n_v)) + n_v \cdot \frac{1}{n_v} - n_v \cdot \frac{1}{N + n_v} \right]$$

$$= \Delta H_v - T\Delta S_v - RT \ln \frac{n_v}{N + n_v} = 0 \tag{14.33}$$

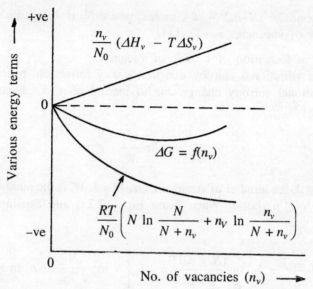

Fig. 14.6 Various terms of Eq. (14.32) as function of number of vacancies at constant temperature (schematic).

Upon rearrangement,

$$X_v(\text{eq}) = \frac{n_v(\text{eq})}{N + n_v(\text{eq})} = \exp\left(\frac{\Delta S_v}{R}\right)\exp\left(-\frac{\Delta H_v}{RT}\right) = \exp\left(-\frac{\Delta G_v}{RT}\right) \qquad (14.34)$$

where ΔG_v = free energy of formation of 1 mole of vacancy and, $X_v(\text{eq})$ = equilibrium vacancy fraction in lattice at temperature T.

Equilibrium concentration of interstitials can also be arrived at following the procedure adopted above, and, in analogy with Eq. (14.34), we may write

$$X_i(\text{eq}) = \frac{n_i(\text{eq})}{N + n_i(\text{eq})} = \exp\left(\frac{\Delta S_i}{R}\right)\exp\left(-\frac{\Delta H_i}{RT}\right) = \exp\left(-\frac{\Delta G_i}{RT}\right) \qquad (14.35)$$

where the subscript i denotes interstitial.

Additional comments

Equations (14.34) and (14.35) are applicable to vacancies and interstitials originating from thermal motions of atoms. These are known as *thermal or intrinsic* vacancies and interstitials. There is another type, which is fixed by composition and is independent of temperature, i.e. the *extrinsic* type. In metals, the best known example is carbon dissolved in iron as extrinsic interstitial. The carbon atoms are located at interstitial sites in iron lattice due to their small atomic radii.

ΔS_v, ΔS_i values are very small due to very low concentrations of vacancies and interstitials. Hence, Eqs. (14.34) and (14.35) may be approximated as

$$X_v(\text{eq}) = \exp\left(-\frac{\Delta H_v}{RT}\right) \tag{14.36}$$

$$X_i(\text{eq}) = \exp\left(-\frac{\Delta H_i}{RT}\right) \tag{14.37}$$

ΔH_v and ΔH_i were calculated theoretically for materials like copper. It was found that for copper, $\Delta H_i/\Delta H_v$ is approximately 7. At 1000 K, from Eqs. (14.36) and (14.37),

$$\frac{X_v(\text{eq})}{X_i(\text{eq})} \cong 10^{39}$$

From this result we may make a generalized statement that, for a close-packed crystal such as copper, the concentration of interstitials is negligible as compared to that of vacancies.

It was also calculated that, for Cu at 1000 K, $X_v(\text{eq}) \cong 3 \times 10^{-7}$ This is indeed small. Hence, vacancies will obey Henry's Law, and all properties related to the same will be proportional to vacancy concentration. One such property is change of electrical resistivity ($\Delta\rho$) of a metal due to vacancies. From Eq. (14.36), we obtain

$$\frac{\delta[\ln X_v(\text{eq})]}{\delta\left(\frac{1}{T}\right)} = -\frac{\Delta H_v}{R} = \frac{\delta\ln(\Delta\rho)}{\delta\left(\frac{1}{T}\right)} \tag{14.38}$$

Figure 14.7 presents $\Delta\rho$ for gold as function of $1/T$. From the slope, ΔH_v was determined as 92.5 kilojoules per mole of vacancy. For some other metals also, the values are near about this. These values also agree reasonably with theoretically calculated ones.

14.3.3 Point Defects in Ionic Compounds

Amongst ionic compounds, the oxides have been most widely studied in view of the fact that their importance is next only to metals and alloys in metallurgy and materials science.

Compounds may be classified into stoichiometric compounds and non-stoichiometric compounds.

(i) *Stoichiometric compounds.* Examples are SiO_2, Al_2O_3, ZrO_2, CaO, ZnO, where metal/oxygen ratios are simple as in their chemical symbols,

(ii) *Nonstoichiometric compounds.* Fe_xO is an example, where $0.85 < x < 0.97$, and Fe/O ratio is thus not a simple one. Another example is VO_x.

In stoichiometric compounds, point defects, as mentioned in section 14.3.2, originate either from composition effect or due to thermal disturbances. In ionic crystals, there are two sub-lattices, viz. anionic and cationic. An anion cannot occupy a cationic sub-lattice, and vice versa, since these require very high energies. Moreover, neighbouring positive charges and negative charges have to be equal for local charge neutrality.

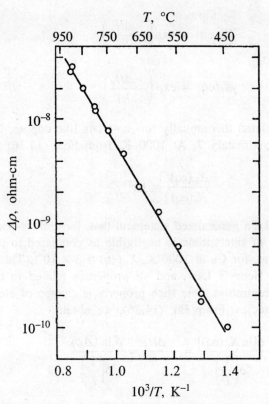

Fig. 14.7 Change of electrical resistivity of quenched gold wires as function of quench temperature [J.E. Bauerie and J.S. Kochler, *Phys. Rev.*, **107**, 1493, 1957].

Stoichiometric compounds

In Section 13.1, we have discussed ZrO_2—CaO solid electrolyte. It was mentioned there that, when Ca^{2+} replaces Zr^{4+} in cationic sub-lattice, two + charges are less in the cationic site occupied by Ca^{2+}. For local charge neutrality, an adjacent O^{2-} ion site remains vacant, thus creating O^{2-} vacancy. This is an example of a stoichiometric compound where the point defects are generated by composition effect.

Point defects arising from thermal effects will always occur in pairs—one positive and another negative. These may be vacancy-interstitial pair, vacancy-vacancy pair, or pairs where one is an electronic defect (excess electron or electron hole). For example, NaCl and KCl exhibit *Schottky defects,* i.e. equal number of vacancies in M sub-lattice and X sub-lattice, where M and X denote metal and nonmetal respectively.

As an example, let us consider a crystal MX, where M^+ and X^- are singly charged. Then the reaction generating Schottky defects may be written as

$$\text{MX} = M_{(1-\delta)}X_{(1-\delta)} + \delta V_M + \delta V_X \qquad (14.39)$$

The activity of MX is 1 for pure MX. Since $\delta \ll 1$, the activity of $M_{(1-\delta)}$, $X_{(1-\delta)}$ also may be set equal to 1. Therefore, the equilibrium relation for reaction (14.39) may be expressed as

$$K_{39} = [V_M][V_X] \tag{14.40}$$

where $[V_M]$ and $[V_X]$ are concentrations of vacancies.

Again, for the reaction

$$V_M = V_M^- + e^+ \tag{14.41}$$

we have

$$K_{41} = \frac{\lfloor e^+ \rfloor \lfloor V_M^- \rfloor}{[V_M]} \tag{14.42}$$

The basis of Eq. (14.41) is that a vacancy in cationic sub-lattice is equivalent to a negative charge. e^+ denotes an electron hole. Similarly, for

$$V_X = V_X^+ + e^- \tag{14.43}$$

$$K_{43} = \frac{\lfloor e^- \rfloor \lfloor V_X^+ \rfloor}{[V_X]} \tag{14.44}$$

In addition to the above, further relations are obtained from conditions of electroneutrality, stoichiometry and relation between $[e^+]$ and $[e^-]$, as follows:

$$\lfloor V_M^- \rfloor + \lfloor e^- \rfloor = \lfloor V_X^+ \rfloor + \lfloor e^+ \rfloor \tag{14.45}$$

$$[V_M] + \lfloor V_M^- \rfloor = [V_X] + \lfloor V_X^+ \rfloor \tag{14.46}$$

$$K_{47} = [e^+][e^-] \tag{14.47}$$

In principle, the six unknown concentrations, viz. $[V_M]$, $\lfloor V_M^- \rfloor$, $[V_X]$, $\lfloor V_X^+ \rfloor$, $[e^-]$ and $[e^+]$ can be solved from the four equilibrium relations and two charge and stoichiometric balances.

Non-stoichiometric compounds

Let us take the example of Wustite (Fe_xO) again. In section 11.2.2, we discussed general issues related to the activities of Fe, O, "FeO" in the Wustite field and how these are related to partial pressure of oxygen in the surrounding atmosphere, with which it was assumed to be at equilibrium. To sum up, the value of x would be governed both by temperature and p_{O_2}. Therefore, defect concentrations in Wustite also will depend on the above variables. Experiments have established that the dominant point defect is cation vacancy. Moreover, it is a p-type semiconductor, i.e. it has electron holes. Some other examples of this type of defect are NiO, Cu_2O, FeS. Figure 14.8 presents log χ vs. log p_{O_2} in Wustite at 1223 and 1273 K. From the slope, it is found that log $\chi \propto p_{O_2}^{1/6}$ approximately, where χ is electrical conductivity. Recognizing that χ is proportional to concentration of electron holes, $\lfloor e^+ \rfloor \propto p_{O_2}^{1/6}$.

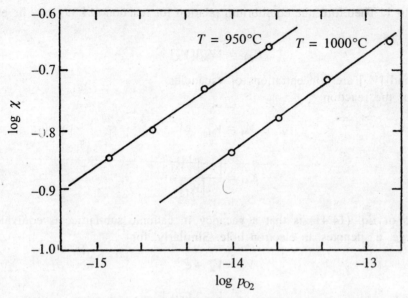

Fig. 14.8 Dependence of electrical conductivity of Wustite on oxygen pressure [K. Hauffe and H. Pfeiffer, *Z. Metallik*, **44**, 27, 1953].

A thermodynamic analysis of the above is as follows:

$$\frac{1}{2}O_2(g) = O_O + 2e^+ + V_{Fe}^{2-} \qquad (14.48)$$

where O_O is oxygen in lattice site, V_{Fe}^{2-} is vacancy in Fe^{2+} site, and hence is equivalent to two negative charges. Since O_O is very large compared to defect concentration, it hardly changes and can be assumed as constant. Hence, we may write

$$K_{48} = \frac{[e^+]^2[V_{Fe}^{2-}]}{p_{O_2}^{1/2}} \qquad (14.49)$$

From charge balance,

$$[e^+] = 2[V_{Fe}^{2-}] \qquad (14.50)$$

Combining Eqs. (14.49) and (14.50), we get

$$[e^+]^3 \propto p_{pO_2}^{1/2}, \text{ i.e } [e^+] \propto p_{pO_2}^{1/6} \qquad (14.51)$$

14.4 Summary

1. A surface possesses some extra energy due to the work done to create it. The relationship is

$$-\delta W' = dG'_s = \sigma \, dA$$

where G'_s is excess surface free energy, σ is specific surface energy per unit surface area, and A is surface area.

2. σ is lowered by the phenomenon of adsorption at the surface. The adsorption equilibrium relations at constant temperature are known as *adsorption isotherms*. We are primarily concerned with chemisorption. For this, the following two are standard isotherms:

   (i) The Langmuir adsorption isotherm deals with monolayer adsorption of ideal gases on a surface.

   (ii) The Gibbs adsorption isotherm deals with adsorption of a component of a solution onto its surface, and is given as

$$\Gamma_i = - \frac{d\sigma}{d\mu_i}$$

where Γ_i is known as surface excess, and μ_i is chemical potential of component i in the bulk solution. The application of this isotherm has been discussed for chemisorption of oxygen dissolved in liquid iron, onto its surface.

3. A phase of a small size has high surface area-to-volume ratio, and hence, a high surface energy per unit volume. This causes excess pressure in small gas bubbles in liquid, higher equilibrium vapour pressure of a liquid droplet, and change of phase transformation temperatures for fine particles. The appropriate equations have been derived in this chapter.

4. In crystalline metallic solids, the defects are surfaces/interfaces, line defects and point defects (vacancies and interstitials). Brief discussions on thermodynamics of grain boundary, which is a special type of interface, and vacancies and interstitials have been given in this chapter.

5. We have given derivations for dependence of equilibrium concentrations of vacancies [$X_v(\text{eq})$] and interstitials [$X_i(\text{eq})$] on temperature, i.e.

$$X_v(\text{eq}) = \exp\left(- \frac{\Delta H_v}{RT}\right)$$

$$X_i(\text{eq}) = \exp\left(- \frac{\Delta H_i}{RT}\right)$$

where ΔH_V and ΔH_i are enthalpies of formation per mole of vacancies and interstitials, respectively. We have also shown that $X_V(\text{eq}) \gg X_i(\text{eq})$.

6. In ionic compounds, all the above defects are present. However, additional features arise due to electrical charges associated with vacancies and interstitials, as well as existence of electronic defects. In nonstoichiometric oxides, again, defect concentrations depend on oxygen potential. In this chapter we have presented some thermodynamic derivations.

Appendix

Selected Thermodynamic Data

Table A-1 Molar Heat Capacities at Constant Pressure

Substance	$C_P = a + bT + cT^{-2}$, joules/mol/K			
	a	$b \times 10^3$	$c \times 10^{-5}$	Range, K
Al(s)	20.7	12.4		298–932
Al(l)	29.0			932–1273
Al_2O_3(s)	106.6	17.8	–28.5	298–1800
C(dia)	9.12	13.2	–6.19	298–1200
C(gr)	17.2	4.27	–8.79	298–2300
CO(g)	28.42	4.10		298–2500
CO_2(g)	44.16	9.04		298–2500
Cr(s)	24.4	9.87	–3.7	$298–T_m$
Cr_2O_3(s)	119.4	9.2	–15.6	350–1800
FeO(s)	48.4	8.37		298–1200
Fe_3O_4(s)	91.6	201.8		298–900
Mn(α)	21.6	15.9		298–993
Mn(β)	34.9	2.8		993–1373
Mn(γ)	44.8			1373–1409
Mn(δ)	47.3			1409–1517
Mn(l)	46.0			$1517–T_b$
MnO(s)	46.48	8.12	–3.7	298–1800
N_2(g)	27.88	4.27		298–2500
Ni(s)	32.6	–1.97	–5.59	298–630
NiO(s)	–20.9	157.2	16.3	298–525

(Contd.)

Table A-1   (Contd.)

Substance	$C_P = a + bT + cT^{-2}$, joules/mol/K			
	a	$b \times 10^3$	$c \times 10^{-5}$	Range, K
$O_2(g)$	30.0	4.18	−1.7	298–3000
$Pb(s)$	23.6	9.75		298–600
$Ti(\alpha)$	22.09	10.04		298–1155
$Ti(\beta)$	28.91			1155–1350
$TiC(s)$	49.5	3.35	14.98	298–1800
$Zn(s)$	22.4	10.0		298–692
$Zn(l)$	31.4			692–1180
$Zn(g)$	20.8			1180–T_b
$ZnO(s)$	49.02	5.11		298–1600

Table A-2   Standard Molar Heats of Formation and Molar Entropies at 298 K

Substance	ΔH_f^0 (298), joules	S_{298}^0, joules/K
$Al(s)$	0	28.34
$Al_2O_3(s)$	−1,674,000	51.1
$C(diamond)$	1,900	2.44
$C(graphite)$	0	5.7
$CO(g)$	−111,700	198.0
$CO_2(g)$	−394,100	213.9
$Cr(s)$	0	23.78
$Cr_2O_3(s)$	−1,120,300	81.2
$FeO(s)$	−259,600	58.81
$Fe_3O_4(s)$	−1,091,000	151.53
$MnO(s)$	−384,700	59.12
$Ni(s)$	0	29.8
$NiO(s)$	−244,600	38.0
$O_2(g)$	0	205.1
$Ti(\alpha)$	0	30.56
$TiC(s)$	−183,700	24.3
$W(s)$	0	33.48
$WO_2(s)$	−589,800	66.95
$WO_3(s)$	−844,300	83.3
$ZnO(s)$	−348,300	43.53

Table A-3 Molar Heats of Fusion and Transformations

Substance	Transformation	ΔH^0_{trans}, joules	T_{trans}, K
Al	s → l	10,500	932
Bi	s → l	10,900	544
Cr	s → l	21,000	2173
Cu	s → l	12,970	1356
Mn	α → β	2,010	993
Mn	β → γ	2,300	1373
Mn	γ → δ	1,800	1409
Mn	δ → l	13,400	1517
Pb	s → l	4,810	600
Si	s → l	50,630	1683
Ti	α → β	3,473	1155
Zn	s → l	7,388	692
Zn	l → g	114,280	1180

Table A-4 Standard Free Energy Changes for Some Reactions

Reaction	ΔG^0, joules	Range, K
$2Al(l) + 3/2\ O_2(g) = Al_2O_3(s)$	$-1,676,000 + 320T$	923–1800
$CaCO_3(s) = CaO(s) + CO_2(g)$	$-126,400 + 144T$	449–1500
$C(s) + 2H_2(g) = CH_4(g)$	$-69,120 + 22.25T\ \ln T - 65.34T$	298–1200
$C(s) + \frac{1}{2}O_2(g) = CO(g)$	$-111,700 - 87.65T$	298–2500
$C(s) + O_2(g) = CO_2(g)$	$-394,100 - 0.84T$	298–2000
$2Cr(s) + 3/2\ O_2(g) = Cr_2O_3(s)$	$-1,120,800 + 260T$	298–2100
$2Cu(s) + \frac{1}{2}O_2(g) = Cu_2O(s)$	$-169,000 - 7.12T\ \ln T + 123T$	298–1356
$2Cu(s) + \frac{1}{2}S_2(g) = Cu_2S(s)$	$-142,900 - 11.3T\ \ln T + 120.2T$	623–1360
$Fe(s) + \frac{1}{2}O_2(g) = FeO(s)$	$-259,600 + 62.55T$	298–1642
$Fe(s) + \frac{1}{2}O_2(g) = FeO(l)$	$-232,700 + 45.31T$	1808–2000
$3Fe(s) + 2O_2(g) = Fe_3O_4(s)$	$-1,091,060 + 312.75T$	298–1642
$Fe(s) + \frac{1}{2}S_2(g) = FeS(l)$	$-113,400 + 26.45T$	1465–1809
$H_2(g) + Cl_2(g) = 2HCl(g)$	$-182,200 + 3.60T\ \ln T - 43.68T$	298–2100
$H_2(g) + \frac{1}{2}O_2(g) = H_2O(g)$	$-246,000 + 54.8T$	298–2500
$H_2(g) + \frac{1}{2}S_2(g) = H_2S(g)$	$-90,290 + 49.39T$	298–2000
$Mg(l) + Cl_2(g) = MgCl_2(l)$	$-605,000 + 125.4T$	973–1133
$Mg(g) + \frac{1}{2}O_2(g) = MgO(s)$	$-759,800 - 13.39T\ \ln T + 316.7T$	1380–2500
$Mn(s) + \frac{1}{2}O_2(g) = MnO(s)$	$-384,700 + 72.8T$	298–1500

(Contd.)

Table A-4   (Contd.)

Reaction	ΔG^0, joules	Range, K
$Mn(l) + \tfrac{1}{2}S_2(g) = MnS(l)$	$-262,600 + 64.4T$	1803–2000
$Mo(s) + O_2(g) = MoO_2(s)$	$-547,270 + 143T$	298–1300
$\tfrac{1}{2}O_2(g) = [O]_{1 \text{ wt\% in Fe}}$	$-116,740 - 6.11T$	1823–1923
$Si(l) = [Si]_{1 \text{ wt\% in Fe}}$	$-119,240 - 52.5T$	1823–1923
$3Si(s) + 2N_2(g) = Si_3N_4(s)$	$-741,000 - 10.5T \ln T + 403T$	298–1686
$Si(s) + O_2(g) = SiO_2(s)$	$-902,000 + 174T$	700–1700
$Si(l) + O_2(g) = SiO_2(s)$	$-947,700 + 198.7T$	1700–2000
$Sn(l) + Cl_2(g) = SnCl_2(l)$	$-333,000 + 118T$	520–925
$\tfrac{1}{2}S_2(g) + O_2(g) = SO_2(g)$	$-364,400 + 72.43T$	718–2273
$SO_2(g) + \tfrac{1}{2}O_2(g) = SO_3(g)$	$-94,560 + 89.37T$	318–1800
$Ti(s) + O_2(g) = TiO_2(s)$	$-910,000 + 173T$	298–2080
$Zn(g) + \tfrac{1}{2}O_2(g) = ZnO(s)$	$-482,920 - 18.8T \ln T + 344.7T$	1170–2000
$Zn(g) + \tfrac{1}{2}S_2(g) = ZnS(s)$	$-397,400 - 14.62T \ln T + 313.5T$	1120–2000

Table A-5   Vapour Pressures of Some Metals

$$\ln p(\text{atm}) = -A/T + B \ln T + CT + D$$

Metal	A	B	$C \times 10^3$	D	Range, K
Ag(l)	33,200	-0.85	—	20.31	$T_m - T_b$
Al(l)	37,884	-1.023	—	21.83	$T_m - T_b$
Cu(l)	40,350	-1.21	—	23.7	$T_m - T_b$
Sn(l)	35,697	—	—	12.32	$T_m - T_b$
Zn(l)	15,250	-1.255	—	21.79	693–1180

Table A-6   Atomic Mass of Some Elements

Al	27.0	C	12.0	Cd	112.4
Cr	52.0	Cu	63.5	Fe	55.8
Mn	54.94	O	16.0	Pb	207.2
Si	28.1	Ti	47.9	Zn	65.4

Bibliography

A. TEXTS

Darken, L.S., and Gurry, R.W., *Physical Chemistry of Metals*, McGraw-Hill, New York (1953).

DeHoff, R.T., *Thermodynamics in Materials Science*, McGraw-Hill, New York (1993).

Denbigh, K., *The Principles of Chemical Equilibrium*, Cambridge University Press, Cambridge (1963).

Gaskell, D.R., *Introduction to the Thermodynamics of Materials,* 3rd ed., Taylor & Francis, Washington DC (1995).

Johnson, D.L., and Stracher, G.B., *Thermodynamic Loop Applications in Materials Systems*, The Minerals, Metals and Materials Society, Warrendale, Pennsylvania (1995).

Kubaschewski, O., Spencer, P.J. and Alcock, C.B., *Materials Thermochemistry*, 6th ed., Pergamon Press, Oxford (1993).

Swalin, R.A., *Thermodynamics of Solids*, John Wiley & Sons, New York (1964).

Upadhyaya, G.S., and Dube, R.K., *Problems in Metallurgical Thermodynamics and Kinetics,* Pergamon Press, New York (1977).

B. DATA SOURCES

Elliott, J.F. and Gleiser, M., *Thermochemistry for Steelmaking,* Vol.I, Addison Wesley, Reading, Mass, (1960).

Elliott, J.F., Gleiser, M., and Ramakrishna, V., *Thermochemistry for Steelmaking,* Vol. II, Addison Wesley, Reading, Mass, (1963).

Hultgren, R., Orr, R.L., Anderson, P.D., and Kelley, K.K., *Selected Values of Thermodynamic Properties of Metals and Alloys,* John Wiley & Sons, New York (1963).

JANAF Thermochemical Data, The Dow Chemical Co., Midland, Michigan (1962–63).

Kelley, K.K., *Contributions to the Data on Theoretical Metallurgy XIII*, Bulletin 584, U.S. Govt. Printing Office, Washington DC (1960).

Knacke, O., Kubaschewski, O., and Hesselman, K., *Thermochemical Properties of Inorganic Substances,* Vols. I and II, 2nd ed., Verlag Stahleisen mbH, Dusseldorf, Germany (1991).

Kubschewski, O., Spencer, P.J., and Alcock, C.B., (already noted in TEXTS Section)

Schick, Harold L. *Thermodynamics of Certain Refractory Compounds,* Vol. 2, Academic Press, New York (1966).

The Japan Society for the Promotion of Science, *Steelmaking Data Source Book,* The 19th Committee on Steelmaking, The Gordon and Breach Science Publishers, New York (1984).

Answers to Problems

[*Note:* Unit of energy is either J (i.e. Joules) or kJ (i.e. kilojoules). Wherever in the problem, the amount of substance is not mentioned, the values are per mole (i.e. per gm-mole) of either a substance or per mole of reaction, as the case may be]

CHAPTER 2

1. $\Delta H - \Delta U = -6424$ J
2. (i) $W = -2331$ J, $q = -1639$ J
   (ii) $\Delta U = 692$ J, $\Delta H = 1153$ J
3. (i) $W = q = 11825$ J, $\Delta U = \Delta H = 0$ for 1st stage
   (ii) $W = 13317$ J, $q = 0$, $\Delta U = -13317$ J, $\Delta H = -18800$ J
4. (a) $W = 2748$ J, $q = 0$, $\Delta U = -2748$ J, $\Delta H = -3848$ J
   (b) $W = 4539$ J, $q = \Delta H = 15888$ J, $\Delta U = 11349$ J
5. W(step iii) = 4.51 kJ, q(for cycle) = -8.5 kJ

CHAPTER 3

1. 59.9 kJ
2. -42.2 kJ
3. -186.7 kJ
4. -404.9 kJ
5. $\Delta H° = -288.2$ kJ, $\Delta U° = -284.0$ kJ
6. Process (ii) releases 906 J more heat.
7. -562.5 kJ
8. 1589 K
9. 349.4 kJ

CHAPTER 4

1. ΔS(syst) = -53.26 J/K, ΔS(surr) = 625 J/K
2. (a) 18.1 J/K; (b) 0; (c) -28.72 J/K

3. -171.6 J/K
4. -13.64 J/K
5. $\Delta S(\text{syst}) = -11.64$ J/K, $\Delta S(\text{surr}) = 12.84$ J/K
6. $\Delta S^0(\text{reaction}) = \Delta S(\text{syst}) = -89.39$ J/K, $\Delta S(\text{universe}) = \Delta S(\text{syst} + \text{surr}) = 198.8$ J/K
7. (a) 3589 K;   (b) 0.0075 C_P;   (c) process spontaneous
8. 118.0 J/K.

CHAPTER 6

1. -1729 J
2. 8586 J
3. 16.46 J, No
4. -2.42 K
5. 78450 atm
6. (i) 260.6 kJ;   (ii) 7.07 J/K
7. 908 atm; as the triple point pressure is 5.14 atm, the 1-atm isobar does not pass through the liquid field.
8. 99.95%.

CHAPTER 7

1. 8.5×10^8
2. -2707 J
3. (i) 3.0 J/K;   (ii) -131.1 J
4. $\Delta H^0 = -242.7$ kJ, $\Delta U^\circ = -240.6$ kJ, $\Delta S^\circ = -89.73$ J/K, $\Delta G^\circ = -197.8$ kJ
5. $\Delta H^0 = -602.4$ kJ, $\Delta S^0 = -357.2$ J/K
6. Ni shall not be oxidized; $\Delta G_f^\circ (\text{NiO}) = 108.8$ kJ,
   $\Delta G_f^\circ (Cr_2O_3) = -171.5$ kJ per mole of O_2
7. $Cr_{23}C_6$ stabler than Cr
8. $\Delta H^\circ = 30.42$ kJ; $\Delta S^\circ = 65.9$ J/K; $\Delta G^\circ = 7.35$ kJ
9. (i) 7.1×10^7, 6.4×10^5;   (ii) $\Delta H_f^\circ = -234.4$ kJ, $\Delta S_f^\circ = -84.1$ J/K

CHAPTER 8

1. No oxidation
2. $H_2S \approx 0$, $O_2 = 25$, $H_2O = 50$, $S_2 = 25$ volume %
3. (a) 4.56×10^{-8} atm;   (b) Titanium will get oxidized.
4. 43,800 J heat evolved
5. 0.371
6. 5250 atm
7. (i) 51 %;   (ii) No change
8. Conversion of Cu_2O into Cu_2S possible
9. Not at equilibrium
10. (i) 1166 K;   (ii) 0.053 atm;   (iii) 1.23 atm
    (*Note:* Since 1500 K > 1166 K, no $CaCO_3$ will be present and P_{CO_2} at 1500 K will be governed by Gas Law.)

11. Fe_3O_4 , 836 K
12. MnS will form.
13. $p_{O_2} = 1.06 \times 10^{-5}$ atm, $p_{H_2S} = 0.0151$ atm, $p_{S_2} = 0.356$ atm, $p_{Zn} = 2.61$ atm

CHAPTER 9

1. $\bar{H}_B^m = R\left[\dfrac{\delta(\ln \gamma_B^{\circ})}{\delta\left(\dfrac{1}{T}\right)}\right]_{P,\,com}$, $\bar{V}_B^m = RT\left[\dfrac{\delta(\ln \gamma_B^{\circ})}{\delta P}\right]_{T,\,comp}$

γ_B° = Henry's Law constant, and is not a function of composition.

2. 21000 J, 23.0 J/K
3. (i) $p_{Sn} = 6.3 \times 10^{-7}$ atm, $p_{Cu} = 3.68 \times 10^{-7}$ atm
   (ii) $\gamma_{Sn} = 0.833$, $\gamma_{Cu} = 0.603$
   (iii) $\bar{G}_{Sn}^m = -12800$ J, $\bar{G}_{Cu}^m = -13046$ J
   (iv) $\Delta G^m = -12950$ J, $G^{XS} = -5117$ J
4. $p_{Al} = 2.64 \times 10^{-10}$ atm, $p_{Ag} = 7.1 \times 10^{-6}$ atm; since $p_{Al} \ll p_{Ag}$, removal of Al not feasible.
5. (i) No
   (ii) γ should be constant in Henry's Law region; $X_{Sn} < 0.1$, $X_{Cu} < 0.1$ for validity of HL for Sn, Cu, respectively.
   (iii) and (v) $p_{Sn} = 4.6 \times 10^{-10}$ atm, $p_{Cu} = 4.5 \times 10^{-11}$ atm
   $\bar{G}_{Sn}^m = -4955$ J, $\bar{G}_{Cu}^m = -13922$ J, $\Delta G^m = -8542$ J, $G^{XS} = -2630$ J
   Not regular solution since α_{Sn}, α_{Cu} depend on composition.
6. (ii)

| T, K | Values in J/mole | | | | | |
	$\bar{G}_{Cu}^m$	$\bar{G}_{Ag}^m$	ΔG^m	G^{XS}	α_{Cu}	α_{Ag}
1200	-3558	-3775	-3666	3249	1.346	1.26
1300	-4168	-4491	-4329	3162	1.23	1.11
1400	-4836	-5104	-4970	3097	1.11	1.02

(iii) and (iv) Values in J/mole for H , in J/mole/K for S

$\bar{H}_{Cu}^{XS} = \bar{H}_{Cu}^m = 980$, $\bar{H}_{Ag}^m = 785$, $\Delta H^m = H^{XS} = 882$

$\bar{S}_{Cu}^m = 1.52$, $\bar{S}_{Ag}^m = 1.48$, $\Delta S^m = 1.50$

$\bar{S}_{Cu}^{XS} = -1.36$, $\bar{S}_{Ag}^{XS} = -1.40$, $S^{XS} = -1.38$

(v) Since $S^{XS} \neq 0$, not a regular solution.

7. (ii) $G^{XS} = -4620$ J, $\Delta G^m = -12453$ J
9. (i) $\ln \gamma_{Zn} = 2.00 (1 - X_{Zn})^2 - 0.691 (1 - X_{Zn})^3$; (ii) 3.70; (iii) 506 J

CHAPTER 10

1. μ_C(gas) = -46.24 kJ/mole C ; cannot saturate steel with carbon.
2. $F = 1$; $\mu_{N_2} = -205.3$ kJ; stable phases at 1200 K: Si(s) + gas
3. 3
4. $\mu_{S_2} = -126.3$ kJ; Cu will be converted to Cu_2S.
5. -320.5 kJ
6. $\Delta G_m^\circ = 13010 - 1.3T \ln T + 2.554T$ J
7. (i) 0.003 ppm (i.e. 3 ppb); (ii) Al_4C_3 will form.
8. $\gamma_{Cu}(\beta) = 29.5$

CHAPTER 11

1. (i) 0.396; (ii) p_{CO_2} = 0.049 atm, p_{CO} = 0.951 atm
2. 0.846
3. 1002
4. 1.35×10^{-5}
5. $[X_{Zn}]$ in Pb = 2.4×10^{-5}
6. (i) 2.13×10^{-5} atm; (ii) 1.14×10^{-3} atm
7. (i) X_{FeO} = 0.001, X_{MnO} = 0.999; (ii) 2.6×10^{-27} atm
8. -50.0 kJ per mole $3CaO.Al_2O_3$; 9.79×10^{-3}
9. (i) 1.3; (ii) -163 kJ
10. (i) -36.8 kJ/mole Cu; (ii) -69.6 kJ/mole Mn
11. -0.341
12. 0.0125
13. 0.09 wt.%
14. -567.8 kJ
15. -36.39 kJ

CHAPTER 13

1. ΔG = -314.1 kJ, ΔH = -30.1 kJ, ΔS = 44.1 J/K
2. $-0.022T$ J/mole/K
3. 2.47 volts
4. a_{Al} = 0.673, $\bar{G}_{Al}^m$ = -2152 J, $\bar{H}_{Al}^m$ = 3315 J
5. (i) a_{Sn} = 0.43, $\bar{G}_{Sn}^m$ = -8414 J, $\bar{G}_{Sn}^{XS}$ = -1499 J
   (ii) $\bar{S}_{Sn}^m$ = 11.74 J/K, $\bar{S}_{Sn}^{XS}$ = 5.98 J/K, $\bar{H}_{Sn}^m = \bar{H}_{Sn}^{XS}$ = 5673 J
6. 0.231 volt
7. (i) -127.0 kJ; (ii) -177.6 kJ
8. (i) 0.813; (ii) 0.70

Index